Engineering Drawing for Manufacture

Dedication

To my wife, Jackie, and all my family for being there and the good Lord for allowing us to be here:

> *O Lord…*
> *When I look at thy heavens, the work of thy fingers,*
> *The moon and the stars which thou hast established,*
> *What is man that thou art mindful of him*
> *And the son of man that thou dost care for him?*
> (Psalm 8, verses 3 and 4, the Bible, Revised Standard Version.)

Acknowledgements

The author is grateful to and wishes to acknowledge:

My wife Jackie and all my family for their patience; Lyn Wright and Margaret Hodgeson for their typing and 'Beej' organisation.

To Brian Such for his helpful advice on BSI and ISO matters.

Manufacturing Engineering Modular Series

Engineering Drawing for Manufacture

Brian Griffiths

London and Sterling, VA

First published in Great Britain and the United States in 2003 by Kogan Page Science, an imprint of Kogan Page Limited

120 Pentonville Road
London N1 9JN
UK
www.koganpagescience.com

22883 Quicksilver Drive
Sterling VA 20166–2012
USA

ISBN 1 8571 8033 X

British Library Cataloguing-in-Publication Data

A CIP record for this book is available from the British Library.

Library of Congress Cataloging-in-Publication Data

Griffiths, Brian, 1945-
Engineering drawing for manufacture / Brian Griffiths.
p. cm. – (Manufacturing engineering modular series)
Includes bibliographical references and index.
ISBN 1-85718-033-X
1. Engineering drawings–Standards. I. Title. II. Series.
T352 .G75 2002
604.2′02′18–dc21
2002014373

Typeset by Saxon Graphics Ltd, Derby
Printed and bound in Great Britain by Biddles Ltd, Guildford and King's Lynn
www.biddles.co.uk

Contents

Introduction

In today's global economy, it is quite common for a component to be designed in one country, manufactured in another and assembled in yet another. The processes of manufacture and assembly are based on the communication of engineering information via drawing. These drawings follow rules laid down in national and international standards and codes of practice. The 'highest' standards are the international ones since they allow companies to operate in global markets. The organisation which is responsible for the international rules is the International Standards Organisation (ISO). There are hundreds of ISO standards on engineering drawing and the reason is that drawing is very complicated and accurate transfer of information must be guaranteed. The information contained in an engineering drawing is actually a legal specification, which contractor and subcontractor agree to in a binding contract. The ISO standards are designed to be independent of any one language and thus much symbology is used to overcome a reliance on any language. Companies can only operate efficiently if they can guarantee the correct transmission of engineering design information for manufacturing and assembly.

This book is meant to be a short introduction to the subject of engineering drawing for manufacture. It is only six chapters long and each chapter has the thread of the ISO standards running through it. It should be noted that standards are updated on a five-year rolling programme and therefore students of engineering drawing need to be aware of the latest standards because the goalposts move regularly! Check that books based on standards are less than five years old! A good example of the need to keep abreast of developments is the decimal marker. It is now ISO practice to use a comma rather than a full stop for the decimal marker. Thus, this book is unique in that it introduces the subject of engineering drawing in the context of standards.

The book is divided into six chapters that follow a logical progression. The first chapter gives an overview of the principles of engineering drawing and the important concept that engineering drawing is like a language. It has its own rules and regulation areas and it is only when these are understood and implemented that an engineering drawing becomes a specification. The second chapter deals with the various engineering drawing projection methodologies. The third chapter introduces the concept of the ISO rules governing the representation of parts and features. A practical example is given of the drawing of a small hand vice. The ISO rules are presented in the context of this vice such that it is experiential learning rather than theoretical. The fourth chapter introduces the methods of dimensioning and tolerancing components for manufacture. The fifth chapter introduces the concept of limits, fits and geometric tolerancing, which provides the link of dimensioning to functional performance. A link is also made with respect to the capability of manufacturing processes. The sixth and final chapter covers the methodology of specifying surface finish. A series of questions are given in a final section to aid the students' understanding. Full references are given at the end of each chapter so the students can pursue things further if necessary.

List of Symbols

A	constant
B	constant
f	feed per revolution
m_N	amplitude distribution function moments
Ml(c)	sum of the section lengths
Mr1	upper material ratio
Mr2	lower material ratio
Ra	centre line average
Rdc	height between two section levels of the BAC
Rku	kurtosis
Rmr(c)	material ratio at depth 'c'
Rp	peak height
Rq	RMS average
Rsk	skew
RSm	average peak spacing
Rt	EL peak to valley height
Rv	valley depth
Rz	SL peak to valley height
$R\Delta q$	RMS slope
TnN	general parameter
σ	standard deviation

List of Abbreviations

ADF	amplitude distribution function
ANSI	American National Standards Institute
BAC	bearing area curve
BSI	British Standards Institution
CAD	computer aided design
CDF	cumulative distribution function
CL	centre line
CRS	centres
CSK	countersunk
CYL	cylinder
D	diameter
DIA	diameter
DIN	Deutsches Institut für Normung
DRG	drawing
EDM	electro-discharge machining
EL	evaluation length
GT	geometric tolerance
HEX	hexagonal
ISO	International Standards Organisation
IT	international tolerance
L	lower tolerance limit
MMC	maximum material condition
PCD	pitch circle diameter
R	radius
RAD	radius
RMS	root mean square
SEM	scanning electron microscope
SF	surface finish

SL	sampling length
SP	spherical diameter
SQ	square feature
SR	spherical radius
Sϕ	spherical radius
THD	thread
THK	thick
TOL	tolerance
TPD	Technical Product Documentation
U	upper tolerance limit
VOL	volume
2D	two dimensions
3D	three dimensions
ϕ	diameter
∩	arc

1

Principles of Engineering Drawing

1.0 Introduction

This book is a foundational book for manufacturing engineering students studying the topic of engineering drawing. Engineering drawing is important to manufacturing engineers because they are invariably at the receiving end of a drawing. Designers come up with the overall form and layout of an artefact that will eventually be made. This is the basic object of engineering drawing – to communicate product design and manufacturing information in a reliable and unambiguous manner.

Nowadays, companies operate over several continents. Engineering drawings need to be language-independent so that a designer in one country can specify a product which is then made in another country and probably assembled in yet another. Thus, engineering drawing can be described as a language in its own right because it is transmitting information from the head of the designer to the head of the manufacturer and indeed, the head of the assembler. This is the function of any language. The rules of a language are defined by grammar and spelling. These in turn are defined in grammar books and dictionaries. The language of engineering must be similarly defined by rules that are embodied in the publications of standards organisations. Each country has its own standards organisation. For example, in the UK it is the British Standards Institution (BSI), in the USA it is the American National Standards Institute (ANSI) and in Germany it is the Deutsches Institut für Normung (DIN). However, the most important one is the

International Standards Organisation (ISO), because it is the world's over-arching standards organisation and any company wishing to operate internationally should be using international standards rather than their own domestic ones. Thus, this book gives information on the basics of engineering drawing from the standpoint of the relevant ISO standards. The emphasis is on producing engineering drawings of products for eventual manufacture.

1.1 Technical Product Documentation

Engineering drawing is described as '*Graphical Communications*' in various school and college books. Although both are correct, the more modern term is '*Technical Product Documentation*' (TPD). This is the name given to the whole arena of design communication by the ISO. This term is used because nowadays, information sufficient for the manufacture of a product can be defined in a variety of ways, not only in traditional paper-based drawings. The full title of TPD is '*Technical Product Specification – Methodology, Presentation and Verification*'. This includes the methodology for design implementation, geometrical product specification, graphical representation (engineering drawings, diagrams and three-dimensional modelling), verification (metrology and precision measurement), technical documentation, electronic formats and controls and related tools and equipment.

When the ISO publishes a new standard under the TPD heading, it is given the designation: ISO XXXX:YEAR. The 'XXXX' stands for the number allocated to the standard and the 'YEAR' stands for the year of publication. The standard number bears no relationship to anything; it is effectively selected at random. If a standard has been published before and is updated, the number is the same as the previous number but the 'YEAR' changes to the new year of publication. If it is a new standard it is given a new number. This twofold information enables one to determine the version of a standard and the year in which it was published. When an ISO standard is adopted by the UK, it is given the designation: BS ISO XXXX:YEAR. The BSI has a policy that when any ISO standard is published that is relevant to TPD, it is automatically adopted and therefore rebadged as a British Standard.

In this book the term 'engineering drawing' will be used throughout because this is the term which is most likely to be understood by manufacturing engineering students, for whom the book is written. However, readers should be aware of the fact that the more correct title as far as standards are concerned is TPD.

1.2 The much-loved BS 308

One of the motivating forces for the writing of this book was the demise of the old, much-loved 'BS 308'. This was the British Standard dealing with engineering drawing practice. Many people loved this because it was the standard which defined engineering drawing as applied within the UK. It had been the draughtsman's reference manual since it was first introduced in 1927. It was the first of its kind in the world. It was regularly revised and in 1972 became so large that it was republished in three individual parts. In 1978 a version for schools and colleges was issued, termed 'PD 7308'.

Over the years BS 308 had been revised many times, latterly to take account of the ISO drawing standards. During the 1980s the pace of engineering increased and the number of ISO standards published in engineering drawing increased, which made it difficult to align BS 308 with ISO standards. In 1992, a radical decision was reached by the BSI which was that they would no longer attempt to keep BS 308 aligned but to accept all the ISO drawing standards being published as British Standards. The result was that BS 308 was slowly being eroded and becoming redundant. This is illustrated by the fact that in 1999, I had two 'sets' of standards on my shelves. One was the BS 308 parts 1, 2 and 3 'set', which together summed 260 pages. The other set was an ISO technical drawings standards handbook, in 2 volumes, containing 155 standards, totalling 1496 pages!

Thus, by 1999, it was becoming abundantly clear that the old BS 308 had been overtaken by the ISO output. In the year 2000, BS 308 was withdrawn and replaced by a new standard given the designation BS 8888:2000, which was not a standard but rather a route map which provided a link between the sections covered by the old BS 308 and the appropriate ISO standards. This BS 8888:2000 publication, although useful for guidance between the old BS 308 and the newer ISO standards, is not very user-friendly for students learning the language of engineering drawing. Hence this book was written in an attempt to provide a resource similar to the now-defunct BS 308.

1.3 Drawing as a language

Any language must be defined by a set of rules with regard to such things as sentence construction, grammar and spelling. Different languages have different rules and the rules of one language do not necessarily apply to the rules of another. Take as examples the English and German languages. In English, word order is all important. The subject always comes before the object. Thus the two sentences '*the dog bit the man*' and '*the man bit the dog*' mean very different things. However, in German, the subject and object are defined, not by word order but by the case of the definite or indefinite articles. Although word order is important in German, such that the sequence 'time–manner–place' is usually followed, it can be changed without any loss of meaning. The phrase '*the dog bit the man*' translates to: '*der Hund bisst den Mann*'. The words for dog (Hund) and man (Mann) are both masculine and hence the definite article is 'der'. In this case the man being the object is shown by the change of the definite article to 'den'. Although it may seem strange, the word order can be reversed to: '*den Mann bisst der Hund*' but it still means the dog bit the man. The languages are different but, because the rules are different, clear understanding is achieved. Similar principles apply in engineering drawing in that it relies on the accurate transfer of information via two-dimensional paper or a computer screen. The rules are defined by the various national and/or international standards. The standards define how the shape and form of a component can be represented on an engineering drawing and how the part can be dimensioned and toleranced for manufacture. Thus, it is of no surprise that someone once described engineering drawing as a language.

Despite the fact that there are rules defining a language, whether it be spoken or written, errors can still be made. This is because information, which exists in the brain of person number one is transferred to the brain of person number two. The first diagram in Figure 1.1 illustrates the sequence of information transfer for a spoken language. A concept exists in brain number one that has to be articulated. The concept is thus constrained by the person's knowledge and ability in that language. It is much easier for me to express myself in the English language rather than German. This is because my mother tongue is English whereas I understand enough German to get me across Germany. Thus, knowledge of how to speak a language is a form of noise that can distort communication.

The voice is transmitted through the air which in itself can cause distortions due to, for example, the ambient noise level. This is then received by the ears of the second person and transmitted to the brain. Here there is another opportunity for noise to enter the communication sequence. The game 'Chinese whispers' is based on the fun that you can have as a result of mishearing things. If there is no noise entering the communications sequence, then brain two receives the same concept that brain one wishes to transmit. However, as we all know to our cost, this is not always the case! Perhaps all the above can be summed up by a poster in New York which read, '*I know you believe you understand what you think I said, but I am not sure you realise that what you heard is not what I meant*'!

The same sequence of information transfer applies to drawing (see the second diagram in Figure 1.1). In this case the brain instructs the hands to draw symbols which the receiver's eye observes and transmits to their brain. Again noise can distort the flow of information. Note that this does not depend on language and a design can be transmitted via a drawing even when the two people do not speak the same language. In the case of engineering drawing the symbols are defined by the various ISO standards which are the engineering drawing equivalent of dictionaries and grammar books.

The manner in which a designer draws an artefact can vary. One draughtsman may convey the same information using a different number of views and sections than another. This is termed 'draughtsman's licence'. It is comparable to the way a person may express a thought verbally. By the use of different words and

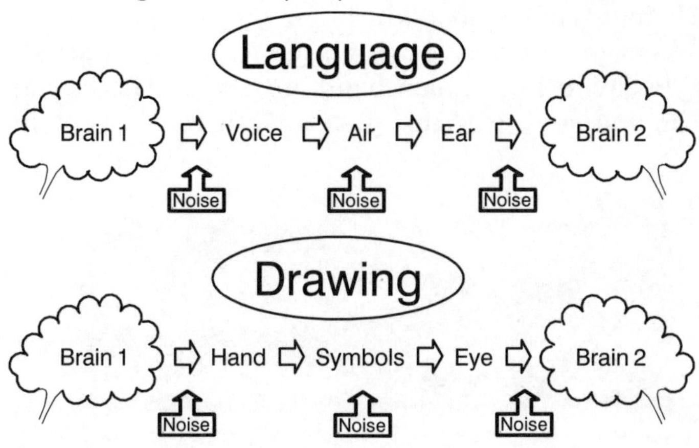

Figure 1.1 *Sources of noise in speech and drawing*

sentences, the same concept can be presented in two or more different ways. Similarly, in engineering drawing, a design may be presented in a variety of ways, all of which can be correct and convey the information for manufacture.

1.4 The danger of visual illusions

Engineering drawing is based on the fact that three-dimensional objects are presented in a two-dimensional form on two-dimensional paper. The potential problems of trying to convey apparent three-dimensional information on two-dimensional flat paper is shown by the two sets of circles in Figure 1.2. The author drew these 12 circles himself, and they are based on a concept by Ramachandran (1988). Because the circles are shaded, each one is seen as either a bump or a depression. In this case, if one's brain interprets the left-hand set of circles as bumps, the right-hand set appears as depressions (and vice versa). During a recent lecture on engineering drawing, I took a vote and two thirds of the student group saw the left-hand set of circles as bumps and the right hand set as depressions. The reason for this is concerned with the shading of the lower part of the circles. Our visual system assumes a single light source. The single light source that we know best is the sun and it shines from above. Thus, the eye sees the left-hand series of circles as bumps because it assumes the illumination is from above. This is not always the case, because in a recent lecture, one third of the students assumed the light source was from below. So much for what the psychologists tell us about the brain!

The facemask in Figure 1.3 is an interesting example of visual illusions (adapted from Ramachandran, 1988). The face appears eerie. Can you guess why this is so without reading any further?

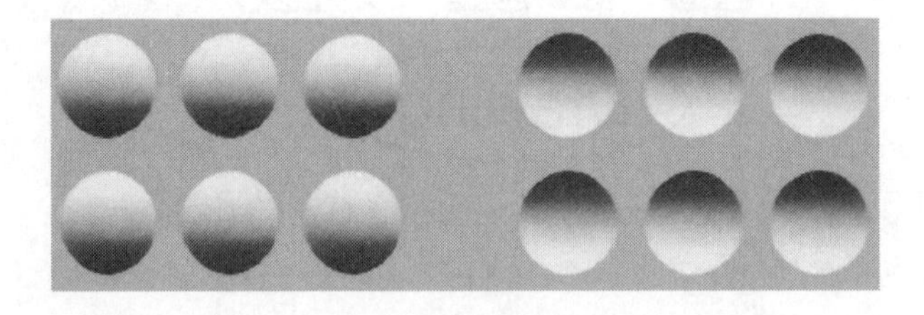

Figure 1.2 *Three-dimensional bumps and depressions*

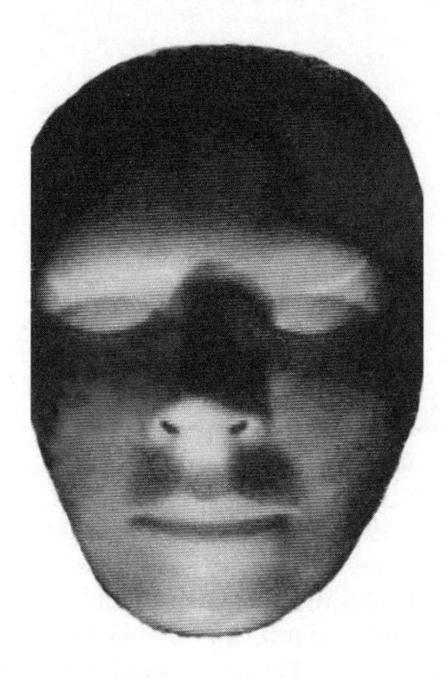

Figure 1.3 *An eerie face mask*

The answer is that it is actually a hollow mask in which the interior is lit from above to produce an eerie impression of a protruding face lit from below. When interpreting shaded images, the brain usually assumes the light is shining from above. Here it rejects that assumption in order to interpret the image as a normal convex object.

The above examples show the difficulties involved in trying to represent three-dimensional information on a two-dimensional piece of paper using shading. A different type of visual illusion is shown in the tri-bar in Figure 1.4. Each of the three corners of the triangle, when considered separately, indicates a valid three-dimensional shape. However, when the tri-bar diagram is considered as an entirety, it becomes an impossible figure. This tri-bar visual illusion was first noted in 1934 by the Swedish artist Oscar Reutersvard. He produced many similar types of drawings of other impossible figures. It was the artist Escher who first bought the knowledge of impossible figures to a much wider audience. He will be particularly remembered for his 'waterfall lithograph' that he produced in 1961. Although channels of water is the subject of his drawing, it is essentially an impossible tri-bar in a different form.

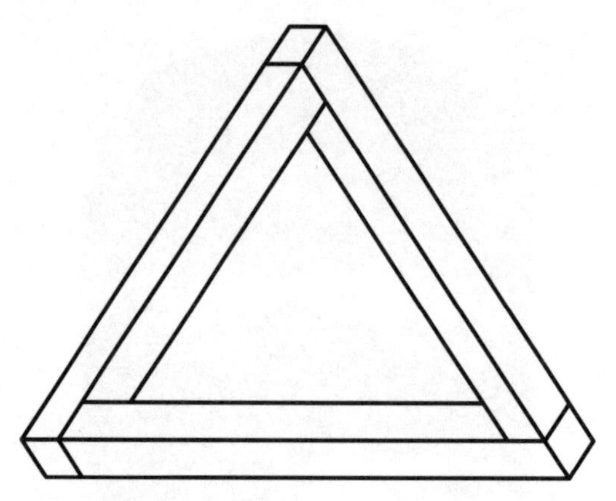

Figure 1.4 *An impossible tri-bar*

The above visual illusions are created because one is trying to represent a three-dimensional object in a two-dimensional space. However, it is still possible to confuse the eye/brain even when absorbing two-dimensional information because rules of perception are broken. The example in Figure 1.5 was actually handed to me on a street corner. The image was written on a credit card-size piece of paper. The accompanying text read, '*Can you find the answer?*'. The problem is that the image breaks one of the pre-conceived rules of perception, which is that the eye normally looks for black information on a white background. In this case the eye sees a jumbled series of shapes and lines. The answer to the question should become obvious when the eye looks for white information on a black background.

Some two-dimensional drawings are termed 'geometrical' illusions because it is the geometric shape and layout that cause distortions. These geometric illusions were discovered in the second half of the 19th century. Three geometric illusions are shown in Figure 1.6. In the 'T' figure, a vertical line and a horizontal line look to be of different lengths yet, in reality, they are exactly the same length. In the figure with the arrows pointing in and out, the horizontal lines look to be of different lengths yet they are equal. In the final figure, the dot is at the mid-point of the horizontal line yet it appears to be off-centre. In all these figures, the eye/brain interprets some parts as different from others. Why this should be so does not seem to be fully understood by psychologists. Gillam (1980/1990) suggests that the effects appear to be related to clues in the size of objects in the three-dimensional world. Although the psychologists

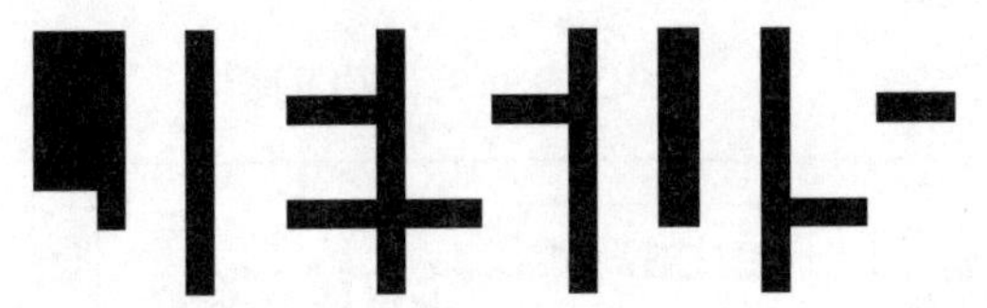

Figure 1.5 *'Can you find the answer?' handout*

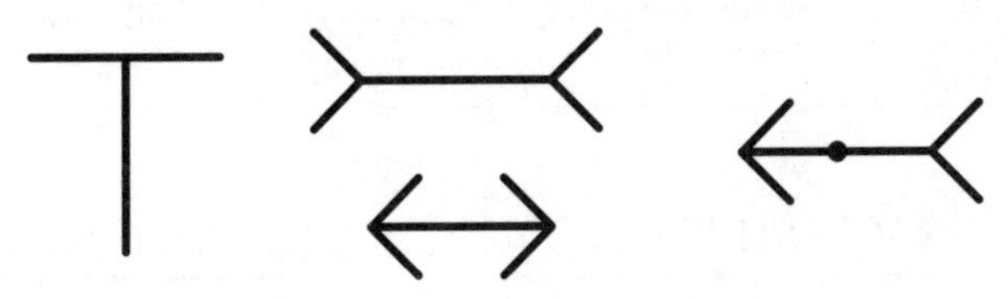

Figure 1.6 *Geometrical optical illusions*

may not understand the theories, the eye/brain sees lines of different length. This is another example of the fact that what the eye sees, even in two dimensions, is not necessarily reality.

All the examples of visual illusions in Figures 1.2 to 1.6 illustrate the complexities involved in firstly representing three-dimensional information in two-dimensional space and secondly making sure the interpretation of the two-dimensional space is correct. It is for these reasons that the fathers of engineering drawing decided that, in orthographic projection, only two-dimensional views would be taken which are projected from one another and things like perspective would be ignored. In this way the 'noise' which could creep into the communication sequence in Figure 1.1 would be reduced to a minimum. The two basic sets of rules of orthographic engineering drawing are based on what is called 'first angle' or 'third angle' projection. The word 'ortho' means right or correct.

1.5 Representation, visualization and specification

1.5.1 Representation and visualization

An artefact or system can be represented in a variety of ways. Engineering drawing is but one of the ways. Figure 1.7 shows some of the ways that products or systems can be represented.

Verbal or written instructions take the form of words describing something. If the words take the form of a set of instructions for doing something, they are ideal. If the words are used to tell a story, then they can paint beautiful pictures in the imagination. However,

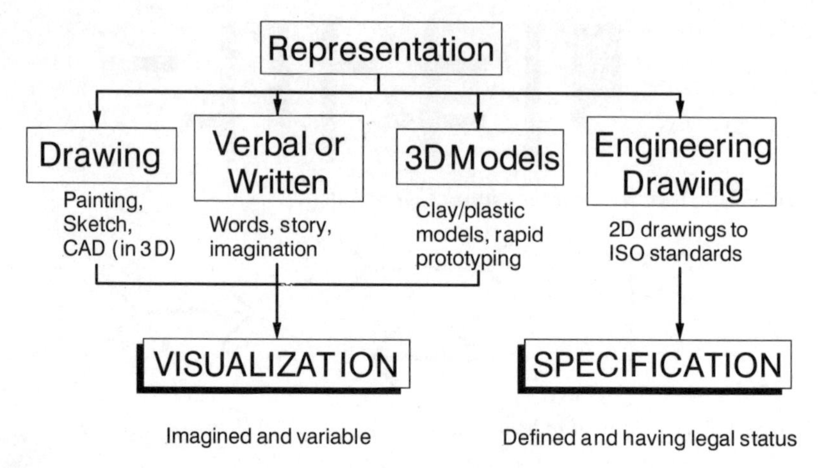

Figure 1.7 *Engineering drawing representation should be specification*

words are clumsy with respect to transmitting information about an engineering artefact. Perhaps a chair or a table could be described without too much difficulty but for anything very much more complex, words become inadequate. Hence, the expression, 'a picture says a thousand words'! Painting or sketching can certainly convey visual information. However, it is also open to artistic interpretation and licence. Ancient pictures of kings and queens often did more credit to them than was justified! Several paintings by Constable of the Dedworth area show the church at different locations because it adds to the artistic balance. Three-dimensional models can certainly be made of engineering artefacts and structures. Indeed, the use of rapid prototyping for the construction of feasibility models is a fast-growing industry. Clay and plastic models have been around for years and mock-ups of new engineering designs for style-based design give the designer a new level of understanding and interpretation. However, three-dimensional models cannot be posted to somebody or sent via the Internet!

All the above techniques can certainly represent an artefact and enable it to be visualized. However, they all fall short of providing a specification that would allow something to be made accurately by a manufacturer speaking a different language to that of the designer. Specification of an artefact is of a higher order than visualization. Specification is needed for engineering artefacts because the instruction to manufacture something given to a subcontractor has financial as well as legal implications. Engineering drawing is a form of engineering representation along with all the others but it is the only one that provides a full specification which allows contracts to be issued and has the support of the law in the 'servant-master' sense. To put it in the words of a BSI drawing manual: '*National and international legal requirements that may place constraints on designs and designers should be identified. ... These may not only be concerned with the aspect of health and safety of material but also the avoidance of danger to persons and property when material is being used, stored, transported or tested*' (Parker, 1991).

1.5.2 Representation and specification

In Section 1.3, it was stated that engineering drawing was the equivalent of a language. A language has to have a set of rules and regulations for it to operate correctly. The same is true of engineering drawing. In the English language, there are two basic rules. The first is the word order that gives information on subject and object. The second is spelling, which gives information on the words themselves in terms of the spelling, i.e. the nouns, verbs, etc. Considering the word order, the phrase '*the cat sat on the mat*' is very different from the phrase '*the mat sat on the cat*'. All that has happened is that the words 'cat' and 'mat' have been swapped. Previously, the phrase described a perfectly feasible situation whereas it now describes an impossible situation. Thus, the word order gives information on which is the subject and which is the object. The second set of rules concern

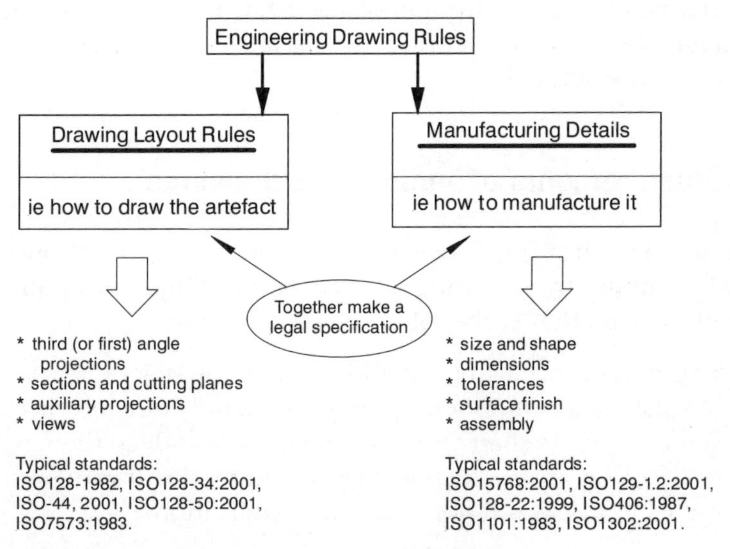

Figure 1.8 *Representation, visualization and specification*

spelling and thus the phrase '*the cat sat on the mat*' is very different from '*the bat sat on the mat*' and yet this difference is the result of only one letter being changed!

In engineering drawing, there are similarly two sets of rules. The first also concerns order but in this case the order of the different orthographic views of an engineering artefact. The second is concerning how the individual views are drawn using different line thicknesses and line types, which is the equivalent of a spelling within each individual word. These are shown in Figure 1.8. The first set are the 'drawing layout rules', which define information concerning the projection method used and therefore the arrangement of the individual views and also the methodology concerning sections. The second set of rules is the 'manufacturing rules', which show how to produce and assemble an artefact. This will be in terms of the size, shape, dimensions, tolerances and surface finish. The drawing layout rules and the manufacturing rules will together make a legal specification that is binding. Both sets of rules are defined by ISO standards. When a contractor uses these two sets of rules to give information to a subcontractor on how to make something, each party is able to operate because of the underpinning provided by ISO standards. Indeed, Chapter 4 will give information concerning a legal argument between a contractor and a subcontractor. The court awarded damages to the subcontractor because the contractor had incorrectly interpreted ISO standards. In another dispute with which the author is familiar, the damages awarded to a contractor because of poor design bankrupted a subcontractor.

1.6 Requirements of engineering drawings

Engineering drawings need to communicate information that is legally binding by providing a specification. Engineering drawings therefore need to met the following requirements:

- Engineering drawings should be unambiguous and clear. For any part of a component there must be only one interpretation. If there is more than one interpretation or indeed there is doubt or fuzziness within the one interpretation, the drawing is incomplete because it will not be a true specification.

- The drawing must be complete. The content of an engineering drawing must provide all the information for that stage of its manufacture. There may be several drawings for several phases of manufacture, e.g. raw shape, bent shape and heat-treated. Although each drawing should be complete in its own right, it may rely on other drawings for complete specification, e.g. detailed drawings and assembly drawings.
- The drawing must be suitable for duplication. A drawing is a specification which needs to be communicated. The information may be communicated electronically or in a hard copy format. The drawing needs to be of a suitable scale for duplicating and of a sufficient scale such that if is micro-copied it can be suitable magnified without loss of quality.
- Drawings must be language-independent. Engineering drawings should not be dependent on any language. Words on a drawing should only be used within the title block or where information of a non-graphical form needs to be given. Thus, there is a trend within ISO to use symbology in place of words.
- Drawings need to conform to standards. The 'highest' standards are the ISO ones that are applicable worldwide. Alternatively standards applicable within countries may be used. Company standards are often produced for very specific industries.

1.6.1 Sizes and layout of drawing sheets

The standard dealing with the sizes and layout of drawing sheets is ISO 5457:1999. If hard copies of drawings are required, the first choice standard sizes of drawings are the conventional 'A' sizes of drawing paper. These sizes are illustrated in Figure 1.9. Drawings can be made in either portrait or landscape orientation but whatever orientation is used, the ratio of the two sides is $1:\sqrt{2}$, (1:1.414). The basic 'A' size is the zero size or '0', known as 'A0'. This has a surface area of $1 m^2$ but follows the $1:\sqrt{2}$ ratio. The relationship is that A1 is half A0, A2 is half A1, etc.

A *blank drawing sheet* should contain the following things (see Figure 1.10). The first three are mandatory, the last four are optional.

1. Title block.
2. Frame for limiting the drawing space.
3. Centring marks.

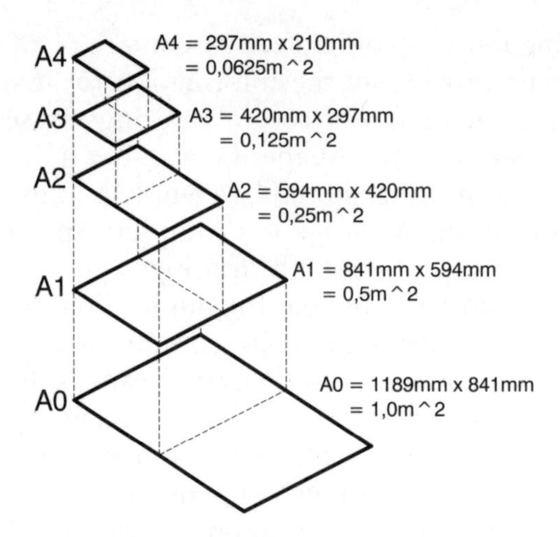

Figure 1.9 *The AO to A4 range of standard drawing sheets sizes*

4. Orientation marks.
5. Metric reference graduation.
6. Grid reference system.
7. Trimming marks.

The *title block* is a specially designated area of the drawing sheet containing information for identification, administration and interpretation of the whole drawing. Irrespective of whether landscape or portrait orientation is used, the title block is normally located in the bottom right-hand corner of the drawing. The information included in the title block can range from the very simple to the exceedingly complex. The manual of British Standards in Engineering Drawing and Design (Parker, 1991) recommends that the following basic information always be included in a title block:

- Name of company or organisation, drawing number, title, date, name of the draughtsman, scale, copyright, projection symbol, measurement units, reference to standards, sheet number, number of sheets and issue information.

The following supplementary information can be provided if necessary:

- Material and specification, heat treatment, surface finish, tolerances, geometrical tolerances, screw thread forms, sheet size, equivalent part, supersedes, superseded by, tool references, gauge references and warning notes.

A *border* should be used to define the edge of the drawing region. It should have a minimum width of 20mm for A0 and A1 sizes and 10mm for A2, A3 and A4. The border shows the edge of the drawing area and would therefore reveal the fact that the drawing had, say, a torn-off corner. The drawing frame is the area within the border (see Figure 1.10).

Trimming marks may be added at the edge of the drawing within the border to facilitate trimming of the paper. There should be four trimming marks at each corner. They can be of two types. The first type is in the form of a right-angled isosceles triangle as shown in the top left-hand corner in Figure 1.10. The second alternative trimming mark is an 'L' shape shown on the top right-hand side of the drawing shown in Figure 1.10.

Centring marks should be provided on the four sides of a drawing to facilitate positioning of the drawing. They take the form of dashes that extend slightly beyond the border as shown in Figure 1.10. They are placed at the centre of each of the four sides.

Orientation marks may be provided on two sides of the drawing sheet (see Figure 1.10). These consist of arrows which coincide with the centring marks. Two such orientation marks should be provided

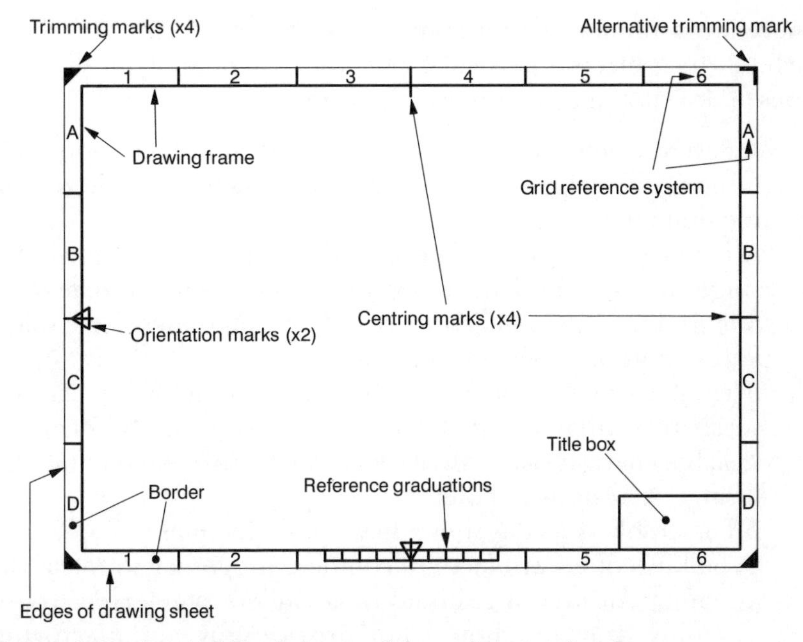

Figure 1.10 *A typical blank sheet used for engineering drawing*

on each drawing, one of which points towards the draughtsman's viewing position.

A *reference metric graduation scale* may be provided with a minimum length of 100mm that is divided into 10mm intervals (see Figure 1.10). The reference graduations consist of 10 off 10mm graduations together making a total length of 100mm. From this graduation scale one can conclude that the drawing size is A3. This calculation shows the usefulness of the reference graduation scale and that it still permits scaling of a drawing when it is presented at a different scale than the original.

An *alphanumeric grid reference system* is recommended for all drawings to permit the easy location of things like details, additions and modifications. The number of divisions should be a multiple of two, the number of which should be chosen with respect to the drawings. Capital letters should be used on one edge and numerals for the other. These should be repeated on the opposite sides of the drawing. ISO 5457:1980 suggests that the length of any one of the reference zones should be not less than 25mm and not more than 75mm.

1.6.2 Types of drawings

There are a number of different types of engineering drawings, each of which meets a particular purpose. There are typically nine types of drawing in common use, these are:

1. A design layout drawing (or design scheme) which represents in broad principles feasible solutions which meet the design requirements.
2. A detail drawing (or single part drawing) shows details of a single artefact and includes all the necessary information required for its manufacture, e.g. the form, dimensions, tolerances, material, finishes and treatments.
3. A tabular drawing shows an artefact or assembly typical of a series of similar things having a common family form but variable characteristics all of which can be presented in tabular form, e.g. a family of bolts.
4. An assembly drawing shows how the individual parts or subassemblies of an artefact are combined together to make the assembly. An item list should be included or referred to. An assembly drawing should not provide any manufacturing

details but merely give details of how the individual parts are to be assembled together.

5. A combined drawing is a combination of detail drawings, assembly drawings and an item list. It represents the constituent details of the artefact parts, how they are manufactured, etc., as well as an assembly drawing and an accompanying item list.
6. An arrangement drawing can be with respect to a finished product or equipment. It shows the arrangement of assemblies and parts. It will include important functional as well as performance requirements features. An installation drawing is a particular variation of an arrangement drawing which provides the necessary details to affect installation of typically chemical equipment.
7. A diagram is a drawing depicting the function of a system, typically electrical, electronic, hydraulic or pneumatic that uses symbology.
8. An item list, sometimes called a parts list, is a list of the component parts required for an assembly. An item list will either be included on an assembly drawing or a separate drawing which the assembly drawing refers to.
9. A drawing list is used when a variety of parts make up an assembly and each separate part or artefact is detailed on a separate drawing. All the drawings and item lists will be cross-reference on a drawing list.

Figures 1.11 and 1.12 show an assembly drawing and a detailed drawing of a small hand vice. The assembly drawing is in orthographic third-angle projection. It shows the layout of the individual parts constituting the assembly. There are actually 14 individual parts in the assembly but several of these are common, such as the four insert screws and two-off hardened inserts such that the number of identifiable separate components numbers 10. On the drawing each of the 10 parts is numbered by a balloon reference system. The accompanying item list shows the part number, the number required and its description. Separate detailed drawings would have to be provided for non-standard parts. One such detailed drawing is shown in Figure 1.12, which is the detailed drawing of the movable jaw. This is shown in third-angle orthographic projection with all the dimensions sufficient for it to be manufactured. Tolerances have been left off for convenience.

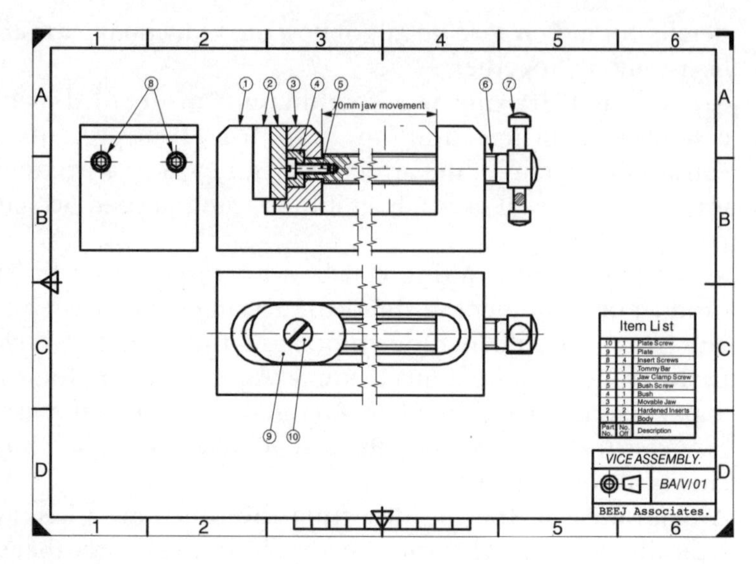

Figure 1.11 *An assembly drawing of a small hand vice*

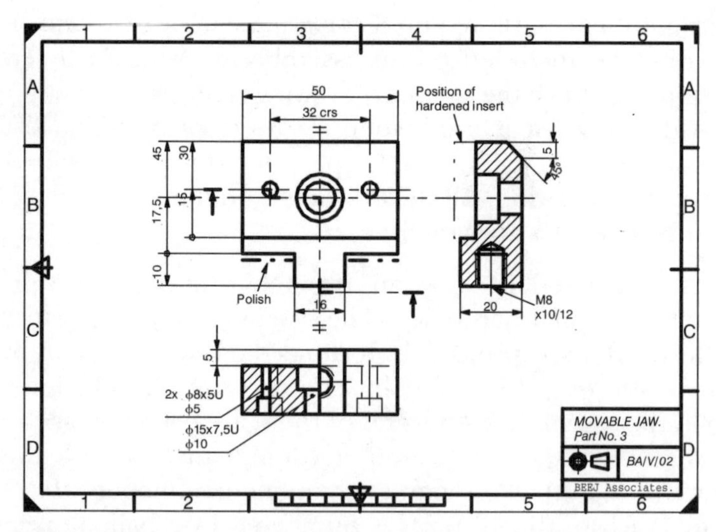

Figure 1.12 *A detailed drawing of the movable jaw of a small hand vice*

1.7 Manual and machine drawing

Drawings can be produced by man or by machine. In the former, it is the scratching of a pencil or pen across a piece of paper whereas in the latter, it is the generation of drawing mechanically via a printer of some type.

In manual drawing, the various lines required to define an artefact are drawn on paper, using draughting equipment. The draughting equipment would typically consist of a surface to draw on, pens or pencils to draw with and aids like set-squares and curves to draw around. A typical drawing surface is a drawing board like the one shown in Figure 1.13 (courtesy of Staedtler). This is a student or lap drawing board with a horizontal ruler that can be moved vertically up and down the board. A small tongue on the left-hand side of the ruler runs in a channel on the side of the board. This allows horizontal lines to be drawn. Rotating the ruler through 90° can allow vertical lines to be drawn. In this position, the ruler tongue runs in the channel running along the bottom of the board (as shown). Alternatively, the arm can be kept in the horizontal position and a draughting head containing an integral set square can be used, which runs in a channel along the centre of the ruler. Often such a drawing head is rotatable and lines can be drawn at any angle. Such drawing boards are typically supplied in A3 and A4 sizes.

The drawing medium can either be a pencil or a pen. Black ink drawing pens are available in a variety of sizes corresponding to the ISO line thicknesses. If pencils are used, clutch pencils are recommended corresponding to the different ISO line thicknesses. A drawing board ruler, like the one described above, enables straight

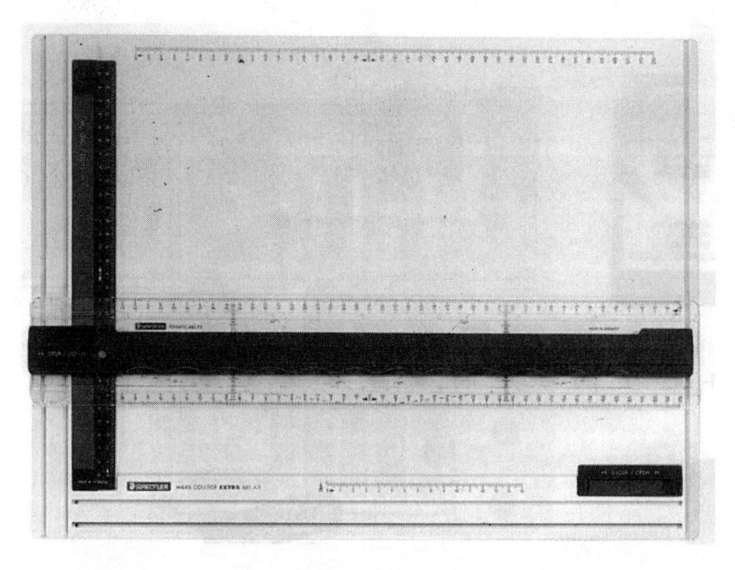

Figure 1.13 *A drawing board (courtesy of Staedtler)*

lines to be drawn but it cannot be used to draw circles. A pair of compasses are used to draw circles. A typical pair of compasses are shown in Figure 1.14 (courtesy of Staedtler). These are fairly expensive 'spring-bow' compasses, so named because the spring ring at the top provides tensioning and allows easy adjustment. Adjustment is achieved by rotating the central thumb wheel. This moves the legs further apart and allows larger diameter circles to be drawn. The compasses shown are pencil compasses that have a stylus point on the left and a pencil lead on the right. In the one shown, the right-hand side pencil leg can be removed and replaced with an ink cartridge pen. Alternative cheaper compasses are available with a simple hinged joint at the top. These are not as convenient to adjust but are more that adequate for everyday needs. Other draughting equipment which is useful but not necessarily mandatory are 'French' curves, flexi-curves, protractors, scaled rulers, lettering stencils and of course the obligatory eraser!

Machine-generated drawings are usually produced on a CAD system. The term 'CAD' is generally assumed to stand for '*computer aided design*' but this is not necessarily the case in engineering drawing. The cheapest CAD systems are really two-dimensional '*computer aided draughting*' packages used on standard PCs. Such a

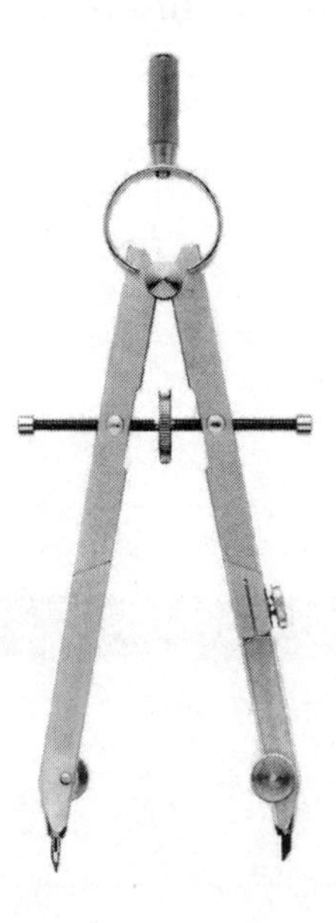

Figure 1.14 *Spring-bow compasses (courtesy of Staedtler)*

two-dimensional draughting package was used to produce the drawings in this book. In this case, the lines are generated on a computer screen using a mouse or equivalent. When the drawing is complete, a printer produces a hard copy on paper. This can be simply plain paper or pre-printed sheets. Systems such as this are limited to two-dimensional drawing in which the computer screen is the equivalent of a piece of paper. True CAD packages are ones in which the computer assists the design process. Such packages can be used to predict stress, strain, deflections, magnetic fields, electrical fields, electrical flow, fluid flow, etc. In integrated CAD packages, artefacts and components are represented in three-dimensions on the two-dimensional screen (called three-dimensional modelling). Parts can be assembled together and modelling can be done to assist the design process. From these three-dimensional models, two-dimensional orthographic engineering drawings can be produced for manufacture.

Irrespective of whether an engineering drawing is produced by manual or machined means, the output for manufacturing purposes is a two-dimensional drawing that conforms to ISO standards. This provides a specification which has a legal status, thus allowing unambiguous manufacture.

References and further reading

BS 308:Part 1:1984, *Engineering Drawing Practice, Part 1, Recommendations for General Principles*, 1984.

BS 308:Part 2:1985, *Engineering Drawing Practice, Part 2, Recommendations for Dimensioning and Tolerance of Size*, 1985.

BS 308:Part 3:1972, *Engineering Drawing Practice, Part 3, Geometric Tolerancing*, 1972.

BS 8888:2000, *Technical Product Documentation – Specification for Defining, Specifying and Graphically Representing Products*, 2000.

Gillam B, 'Geometrical Illusions', *Scientific American*, January 1980. Article in the book *The Perceptual World – Readings from Scientific America Magazine*, edited by Irvin Rock, W H Freeman & Co., 1990.

ISO 5457:1999, *Technical Product Documentation – Sizes and Layouts of Drawing Sheets*, 1999.

ISO Standards Handbook, 'Technical Drawings, Volume 1 – Technical Drawings in General, Mechanical Engineering Drawings & Construction Drawings', third edition, 1977.

ISO Standards Handbook, 'Technical Drawings, Volume 2 – Graphical Symbols, Technical Product Documentation and Drawing Equipment', third edition, 1977.

Parker M (editor), *Manual of British Standards in Engineering Drawing and Design*, second edition, British Standards Institute in association with Stanley Thornes Ltd, 1991.

PD 7308:1986, *Engineering Drawing Practice for Schools and Colleges*, 1986.

Ramachandran V S, 'Perceiving Shape from Shading', *Scientific America*, August 1988. Taken from the book *The Perceptual World – Readings from Scientific America Magazine*, edited by Irvin Rock, W H Freeman & Co., pp 127–138, 1990.

2

Projection Methods

2.0 Introduction

In Chapter 1 it was stated that there are two sets of rules that apply to engineering drawing. Firstly, there are the rules that apply to the layout of a drawing and secondly the rules pertaining to the manufacture of the artefact. This chapter is concerned with the former set of rules, called the 'drawing layout rules'. These define the projection method used to describe the artefact and how the 3D views of it can be represented on 2D paper. These will be presented in terms of first and third angle orthographic projections, sections and cutting planes, auxiliary projections as well as trimetric, dimetric, isometric and oblique projections.

The chart in Figure 2.1 shows the types of drawing projections. All engineering drawings can be divided into either pictorial projections or orthographic projections. The pictorial projections are non-specific but provide visualisation. They can be subdivided further into perspective, axonometric and oblique projections. In pictorial projections, an artefact is represented as it is seen in 3D but on 2D paper. In orthographic projections, an artefact is drawn in 2D on 2D paper. This 2D representation, rather than a 3D representation, makes life very much simpler and reduces confusion. In this 2D case, the representation will lead to a specification that can be defined by laws. The word *ortho* means correct and the word *graphic* means drawing. Thus, *orthographic* means a correct drawing which prevents confusion and therefore can be a true specification which, because orthographic projections are clearly defined by ISO standards, are legal specifications. Orthographic projections can be subdivided into first and third angle projections. The two projection

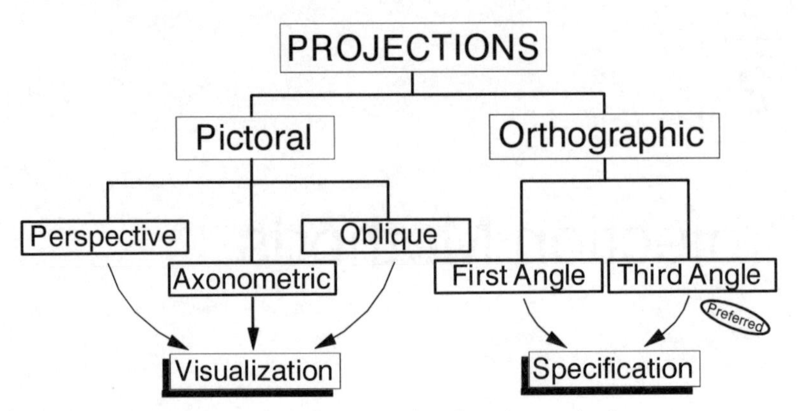

Figure 2.1 *The different types of engineering drawing projections*

methods only differ in the manner in which the views are presented. The third angle projection method is preferred.

Whichever projection method is used, the representation is achieved by projectors which are effectively rays of light whose sources are on one side of an artefact passing over the artefact and projecting its image onto a 2D drawing sheet. This is similar to the image or shadow an artefact would produce when a single light source projects the shadow of an artefact onto, say, a wall. In this case, the wall is the picture plane. The various types of pictorial and orthographic projections are explained in the following sections.

2.1 Perspective projection

Perspective projection is as shown in Figure 2.2. Perspective projection is reality in that everything we see in the world is in perspective such that the objects always have vanishing points. Perspective projection is thus the true view of any object. Hence, we use expressions like 'putting something in perspective'! Projectors radiate from a station point (i.e. the eye) past the object and onto the 2D picture plane. The station point is the viewing point. Although there is only one station point, there are three vanishing points. A good example of a vanishing point is railway lines that appear to meet in the distance. One knows in reality that they never really meet, it is just the perspective of one's viewing point. Although there are three vanishing points, perspective drawings can be simplified such that only two or indeed one vanishing point is used. The drawing in Figure 2.2 shows only two vanishing points.

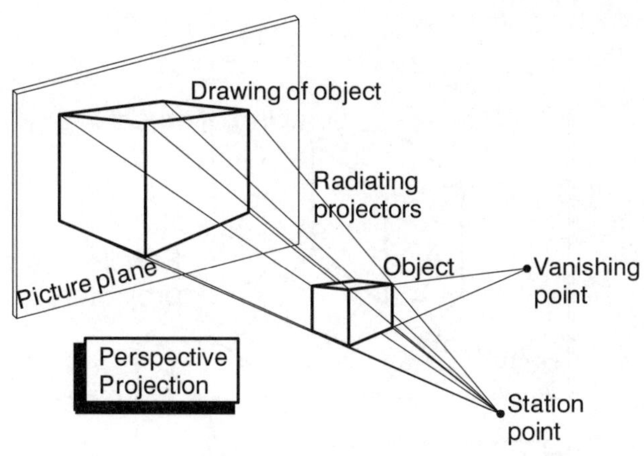

Figure 2.2 *Perspective projection*

Had the block shown been very tall, there would have been a need to have three vanishing points.

Although perspective projection represents reality, it produces complications with respect to the construction of a drawing in that nothing is square and care needs to be taken when constructing such drawings to ensure they are correct. There are numerous books that give details of the methods to be employed to construct perspective drawings. However, for conventional engineering drawing, drawing in perspective is an unnecessary complication and is usually ignored. Thus, perspective projection is very rarely used to draw engineering objects. The problem in perspective projection is due to the single station point that produces radiating projectors. Life is made much simpler when the station point is an infinite distance from the object so that the projectors are parallel. This is a situation for all the axonometric and orthographic projection methods considered below.

2.2 Axonometric projection

Axonometric projection is shown in Figure 2.3. This is the same as perspective projection except that the projectors are parallel. This means that there are no vanishing points. In axonometric projection, the object can be placed at any orientation with respect to the viewer. For convenience, axonometric projection can be divided into three classes depending on the orientation of the object. These are trimetric, dimetric and isometric projections (see Figure 2.4).

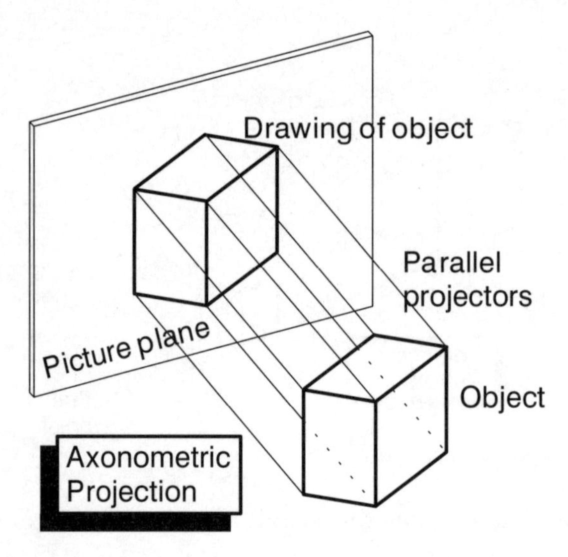

Figure 2.3 *Axonometric projection*

Trimetric projection is by far the most common in that the object is placed at any position with respect to the viewer such that the angles α and β are unequal and the foreshortening in each of the three axes is unequal. The three sides of the cube are of different lengths. This is shown in the left-hand drawing in Figure 2.4. The 'tri' in trimetric means three. In dimetric projection, the angles α and β are the same as shown in the middle drawing in Figure 2.4. This results in equal foreshortening of the two horizontal axes. The third vertical axis is foreshortened to a different amount. The 'di' in dimetric means there are two 'sets' of axes. The particular class of axonometric projection in which all the three axes are foreshortened to an equal amount is called isometric projection. In this case the foreshortening is the same as seen in the right-hand drawing in Figure 2.4. In this case, the angles α and β are the same and equal to 30°. The foreshortening of each of the three axes is identical. The term 'iso' in isometric projection means similar. Isometric projection is the most convenient of the three types of axonometric projection because of the convenience of using 30° angles and equal foreshortening. Isometric projection will be considered in detail in the following section.

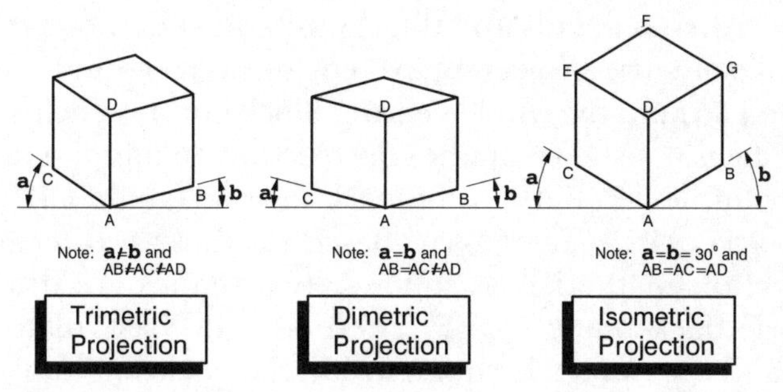

Figure 2.4 *The three types of axonometric projections*

2.3 Isometric projection

In *isometric projection*, the projection plane forms three equal angles with the co-ordinate axis. Thus, considering the isometric cube in Figure 2.4, the three cube axes are foreshortened to the same amount, i.e. AB = AC = AD. Two things result from this, firstly, the angles a = b = 30° and secondly, the rear (hidden) corner of the cube is coincident with the upper corner (corner D). Thus, if the hidden edges of the cube had been shown, there would be dotted lines going from D to F, D to C and D to B. The foreshortening in the three axes is such that AB = AC = AD = $(2/3)^{0.5}$ = 0.816. Since isometric projections are pictorial projections and dimensions are not normally taken from them, size is not really important. Hence, it is easier to ignore the foreshortening and just draw the object full size. This makes the drawing less complicated but it does have the effect of apparently enlarging the object by a factor 1.22 (1 ÷ 0.816). Bearing this in mind and the fact that both angles are 30°, it is not surprising that isometric projection is the most commonly used of the three types of axonometric projection.

The method of constructing isometric projections is shown in the diagrams in Figures 2.5 and 2.6. An object is translated into isometric projection by employing enclosing shapes (typically squares and rectangles) around important features and along the three axes. Considering the isometric cube in Figure 2.4, the three sides are three squares that are 'distorted' into parallelograms, aligned with the three isometric axes. Internal features can be projected from these three parallelograms.

The method of constructing an isometric projection of a flanged bearing block is shown in Figure 2.5. The left-hand drawing shows the construction details and the right-hand side shows the 'cleaned up' final isometric projection. An enclosing rectangular cube could be placed around the whole bearing block but this enclosing rectangular cube is not shown on the construction details diagram because of the complexity. Rather, the back face rectangle CDEF and the bottom face ABCF are shown. Based on these two rectangles, the construction method is as follows. Two shapes are drawn on the isometric back plane CDEF. These are the base plate rectangle CPQF and the isometric circles within the enclosing square LMNO. Two circles are placed within this enclosing square. They represent the outer and inner diameters of the bearing at the back face.

The method of constructing an isometric circle is shown in the example in Figure 2.6. Here a circle of diameter ab is enclosed by the square abcd. This isometric square is then translated onto each face of the isometric cube. The square abcd thus becomes a parallelogram abcd. The method of constructing the isometric circles within these squares is as follows. The isometric square is broken down further into a series of convenient shapes, in this case five small long-thin rectangles in each quadrant. These small rectangles are then translated on to the isometric cube. The intersection heights ef, gh, ij and kl are then projected onto the equivalent rectangles on the isometric projection. The dots corresponding to the points fhjl are the points on the isometric circles. These points can be then joined to produce isometric circles. The isometric

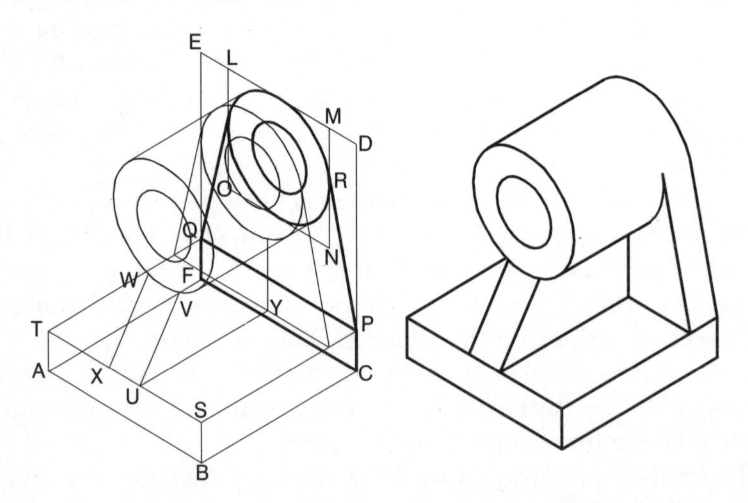

Figure 2.5 *Example of the method of drawing an isometric projection bearing bracket*

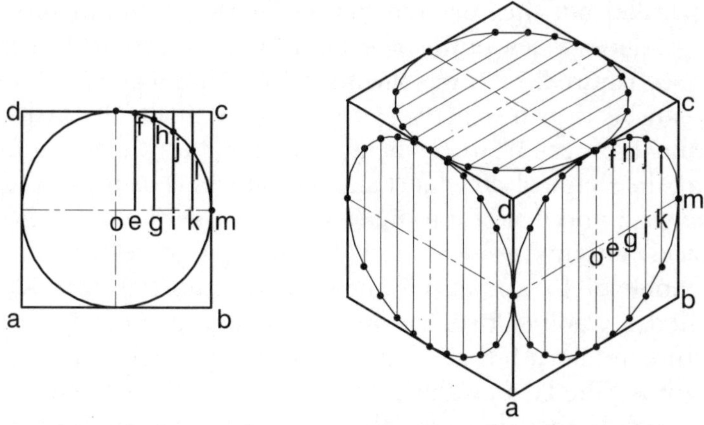

Figure 2.6 *Example of the method of drawing isometric projection circles*

circles can either be produced freehand or by using matching ellipses. Returning to the isometric bearing plate in Figure 2.5, the isometric circles representing the bearing outside and inside diameters are constructed within the isometric square LMNO. Two angled lines PR are drawn connecting the isometric circles to the base CPQF. The rear shape of the bearing bracket is now complete within the enclosing rectangle CDEF.

Returning to the isometric projection drawing of the flanged bearing block in Figure 2.5. The inside and outside bearing diameters in the isometric form are now projected forward and parallel to the axis BC such that two new sets of isometric circles are constructed as shown. The isometric rectangle CPQF is then projected forward, parallel to BC that produces rectangle ABST, thus completing the bottom plate of the bracket. Finally, the web front face UVWX is constructed. This completes the various constructions of the isometric bearing bracket and the final isometric drawing on the right-hand side can be constructed and hidden detail removed.

Any object can be constructed as an isometric drawing provided the above rules of enclosing rectangles and squares are followed which are then projected onto the three isometric planes.

2.4 Oblique projection

In oblique projection, the object is aligned such that one face (the front face) is parallel to the picture plane. The projection lines are still parallel but they are not perpendicular to the picture plane. This produces a view of the object that is 3D. The front face is a true view (see Figure 2.7). It has the advantage that features of the front face can be drawn exactly as they are, with no distortion. The receding faces can be drawn at any angle that is convenient for illustrating the shape of the object and its features. The front face will be a true view, and it is best to make this one the most complicated of the faces. This makes life easier! Most oblique projections are drawn at an angle of 45° and at this angle the foreshortening is 50%. This is called a Cabinet projection. This is because of its use in the furniture industry. If the 45° angle is used and there is no foreshortening it is called a Cavalier projection. The problem with Cavalier projection is that, because there is no foreshortening, it looks peculiar and distorted. Thus, Cabinet projection is the preferred method for constructing an oblique projection.

An oblique drawing of the bearing bracket in Cabinet projection is shown in Figure 2.8. For convenience, the front view with circles was chosen as the true front view. This means that the circles are true circles and therefore easy to draw. The method of construction for oblique projection is similar to the method described above for isometric projection except that the angles are not 30° but 45°. Enclosing rectangles are again used and transposed onto the 45° oblique planes using 50% foreshortening.

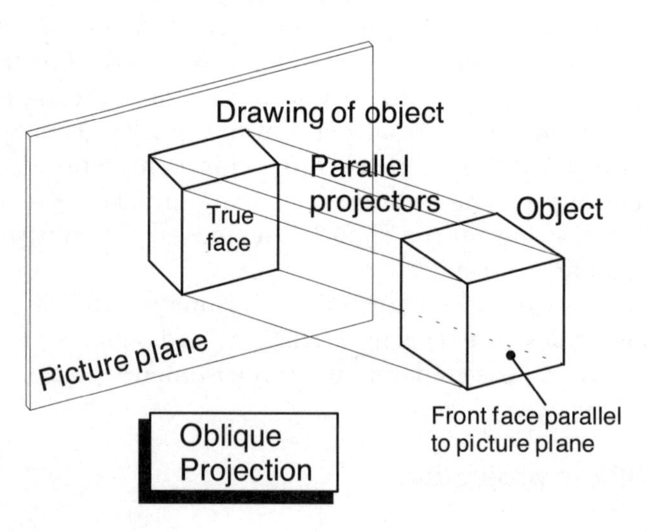

Figure 2.7 *Oblique projection*

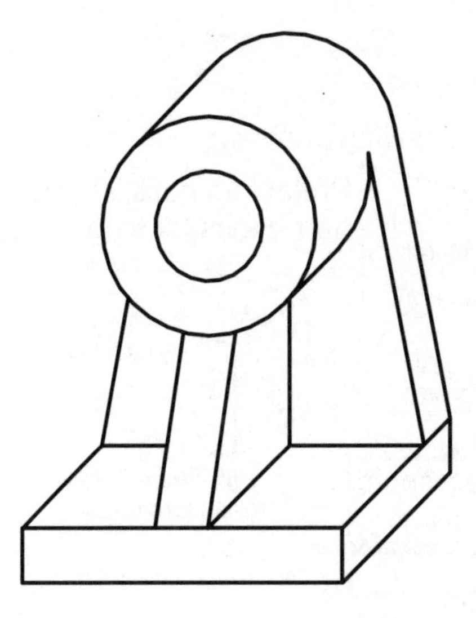

Figure 2.8 *Example of the method of drawing an oblique projection bearing bracket*

2.5 Orthographic projection

In orthographic projection, the front face is always parallel to the picture frame and the projectors are perpendicular to the picture frame (see Figure 2.9). This means that one only ever sees the true front face that is a 2D view of the object. The receding faces are therefore not seen. This is the same as on an oblique projection but with the projectors perpendicular rather than at an angle. The other faces can also be viewed if the object is rotated through 90°. There will be six such orthographic views. These are stand-alone views but if the object is to be 'reassembled' from these six views there must be a law that defines how they are related. In engineering drawing there are two laws, these are first or third angle projection. In both cases, the views are the same; the only thing that differs is the position of the views with respect to each other. The most common type of projection is *third angle projection*.

2.5.1 Third angle projection

Figure 2.10 shows a small cornflake packet (courtesy of Kellogg's) that has been cut and folded back to produce a development of a set of six connected faces. Each one of these faces represents a true view of the original box. Each face (view) is folded out from an adjacent

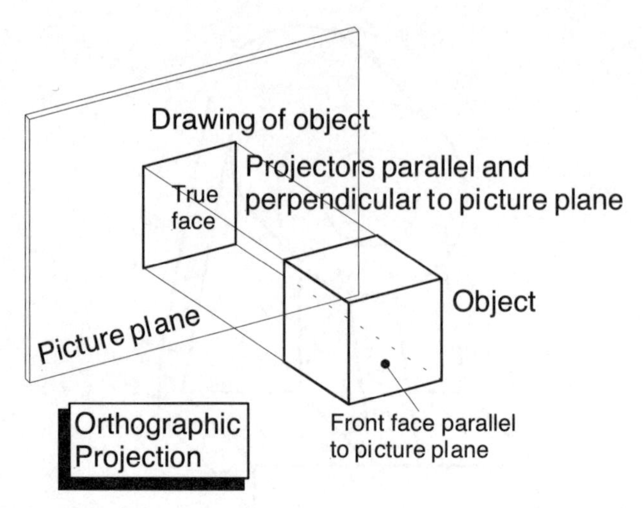

Figure 2.9 *Orthographic projection*

Figure 2.10 *A folded out cardboard cornflake packet (courtesy of Kellogg's)*

face (view). Folding the faces back and gluing could reassemble the packet. The development in Figure 2.10 is but one of a number of possible developments. For example, the top and bottom small faces could have been connected to (projected from) the back face (the 'bowl game' face) rather than as shown. Alternatively, the top and bottom faces could have been connected.

Figure 2.11 (courtesy of Kellogg's) shows the same layout but with the views separated from each other such that it is no longer a development but a series of individual views of the faces. The various views have been labelled. The major face of the packet is the one with the title 'Corn Flakes'. This face is the important one because it is the one that would be placed facing outwards on a supermarket shelf. This view is termed the 'front view' and all the other views are projected from it. Note the obvious names of the other views.

All the other five views are projected from the front face view as per the layout in Figure 2.10. This arrangement of views is called *third angle orthographic projection*. The reason why this is so is explained below. The third angle orthographic projection 'law' is

Figure 2.11 *Cornflake packet in third angle projection (courtesy of Kellogg's)*

that the view one sees from your viewing position is placed on the same side as you view it from. For example, the plan view is seen from above so it is placed above the front face because it is viewed from that direction. The right-side view is placed on the right-hand side of the front view. Similarly, the left-side view is placed to the left of the front view. In this case, the rear view is placed on the left of the left-side view but it could have also been placed to the right of the right-side view. Note that opposite views (of the packet) can only be projected from the same face because orthographic relationships must be maintained. For example, in Figure 2.11, the plan view and inverted plan view are both projected from the front view. They could just as easily have both been projected from the right-side view (say) but not one from the front face and one from the right-side view. It is doesn't matter which arrangement of views is used as long as the principle is followed that you place what you see at the position from which you are looking.

Figure 2.12 shows a third angle projection drawing of a small bracket. In this case, the plan view and the inverted plan view are projected from the front face. Note that the arrangements of the views are still in third angle projection but they are arranged differently from the views in Figure 2.11. Another example of third angle projection is seen in the truncated cone within the title box in Figure 2.12. Here, the cone is on its side and only two views are shown yet they are still in third angle projection. The reason the cone is shown within the box is that it is the standard symbol for third angle

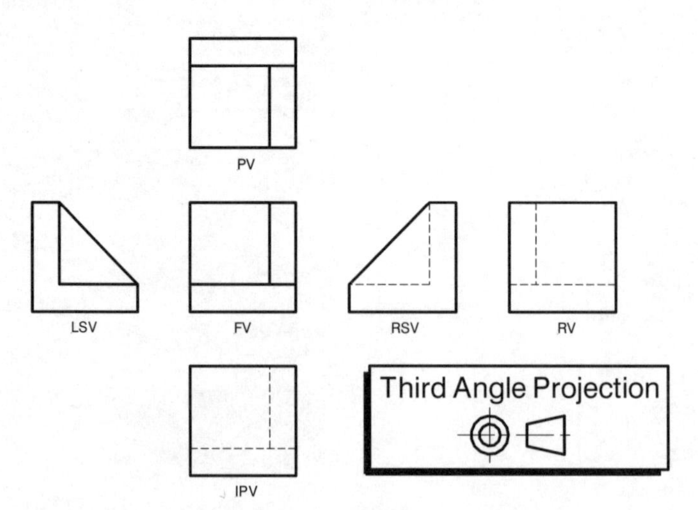

Figure 2.12 *Third angle projection of a bracket*

projection recommended in ISO 128:1982. The standard recommends that this symbol be used within the title block of an engineering drawing rather than the words '*third angle projection*' because ISO uses symbology to get away from a dependency on any particular language.

Third angle projection has been used to describe engineering artefacts from the earliest of times. In the National Railway Museum in York, there is a drawing of George Stephenson's 'Rocket' steam locomotive, dated 1840. The original is in colour. This is a cross between an engineering drawing (as described above) and an artistic sketch. Shadows can be seen in both orthographic views. Presumably this was done to make the drawings as realistic as possible. This is an elegant drawing and nicely illustrates the need for 'engineered' drawings for the manufacture of the Rocket locomotive.

Bailey and Glithero (2000) state, 'The Rocket is also important in representing one of the earliest achievements of mechanical engineering design'. In this context, the use of third angle projection is significant, bearing in mind that the Rocket was designed and manufactured during the transition period between the millwright-based manufacturing practice of the craft era and the factory-based manufacturing practice of the industrial revolution. However, third angle projection was used much earlier than this. It was used by no less than James Watt in 1782 for drawing John Wilkinson's Old Forge engine in Bradley (Boulton and Watt Collection at Birmingham Reference Library). In 1781 Watt did all his own drawing but from 1790 onwards, he established a drawing office and he had one assistant, Mr John Southern.

These drawings from the beginning of the industrial revolution are significant. They illustrate that two of the fathers of the industrial revolution chose to use third angle projection. It would seem that at the beginning of the 18th century third angle was preferred, yet a century later first angle projection (explained below) had become the preferred method in the UK. Indeed, the 1927 BSI drawing standard states that third angle projection is the preferred UK method and third angle projection is the preferred USA method. It is not clear why the UK changed from one to the other. However, what is clear is that it has changed back again because the favoured projection method in the UK is now third angle.

2.5.2 First angle projection

The other standard orthographic projection method is first angle projection. The only difference between first angle and third angle projection is the position of the views. First angle projection is the opposite to third angle projection. The view, which is seen from the side of an object, is placed on the opposite side of that object as if one is looking through it. Figure 2.13 shows the first angle projection layout of the bracket shown in Figure 2.12. The labelling of the views (e.g. front view, plan, etc.) is identical in Figures 2.12 and 2.13. Note that in first angle projection, the right-side view is not placed on the right-hand side of the front view as in third angle projection but rather on the left-hand side of the front view as shown in Figure 2.13. Similarly, the left-side view appears on the right-hand side of the front view. The other views are similarly placed. A comparison between Figures 2.12 and 2.13 shows that the views are identical but the positions and hence relationships are different.

Another first angle projection drawing is seen in the title box in Figure 2.13. This is the truncated cone. It is the standard ISO symbol for first angle projection (ISO 128:1982). It is this symbol which is placed on drawings in preference to the phrase 'first angle projection'.

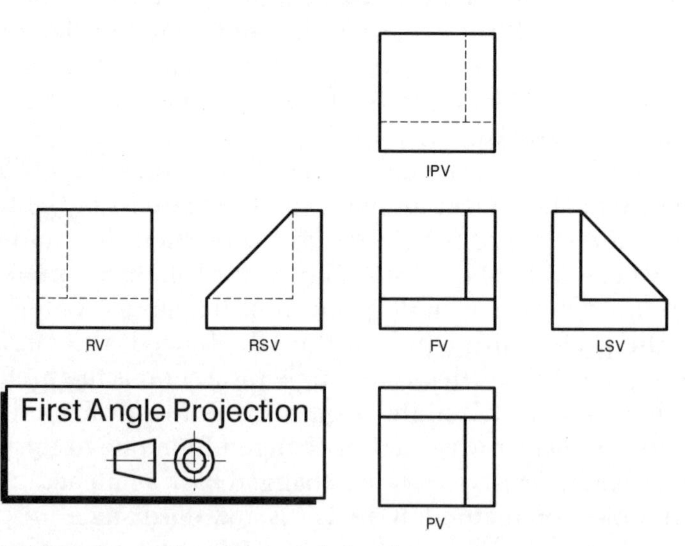

Figure 2.13 *First angle projection of a bracket*

First angle projection is becoming the least preferred of the two types of projection. Therefore, during the remainder of this book, third angle projection conventions will be followed.

2.5.3 Projection lines

In third angle projection, the various views are projected from each other. Each view is of the same size and scale as the neighbouring views from which it is projected. Projection lines are shown in Figure 2.14. Here only three of the Figure 2.12 views are shown. Horizontal projection lines align the front view and the left-side view of the block. Vertical projection lines align the front view and the plan view. The plan view and the left-side view must also be in orthographic third-angle projection alignment but they are not projected directly from one another. A deflector line is placed at 45°. This line allows the horizontal projection lines from the plan view to be rotated through 90° to produce vertical projection lines that align with the left-side view. These horizontal and vertical projection lines are very convenient for aligning the various views and making sure that they are in correct alignment. However, once the views are completed in their correct alignment, the projection lines are not needed because they tend to complicate the drawing with respect to the main purpose, which is to manufacture the artefact.

It is normal industrial practice to erase any projection lines such that the views stand out on their own. Often in engineering drawing

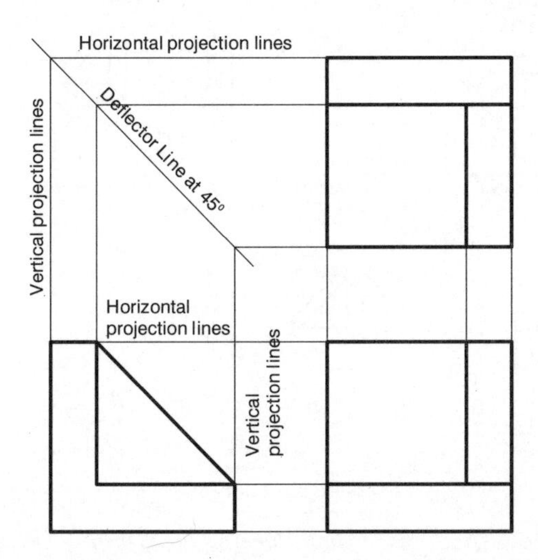

Figure 2.14 *Third angle projection of a bracket showing the projection lines*

lessons in a school, the teacher may insist projection lines be left on an orthographic drawing. This is done because the teacher is concerned about making sure the academic niceties of view alignment are completed correctly. Such projection lines are an unnecessary complication for a manufacturer and therefore, since the emphasis here is on drawing for manufacture, projection lines will not be included from here on in this book.

2.6 Why are first and third angle projections so named?

The terms *first angle projection* and *third angle projection* may seem like complicated terms but the reason for their naming is connected with geometry. Figure 2.15 shows four angles given by the planes OA, OB, OC and OD. When a part is placed in any of the four quadrants, its outline can be projected onto any of the vertical or horizontal planes. These projections are produced by viewing the parts either from the right-hand side or from above as shown by the arrows in the diagram.

In first angle projection the arrows project the shape of the parts onto the planes OA and OB. When the two planes are opened up to

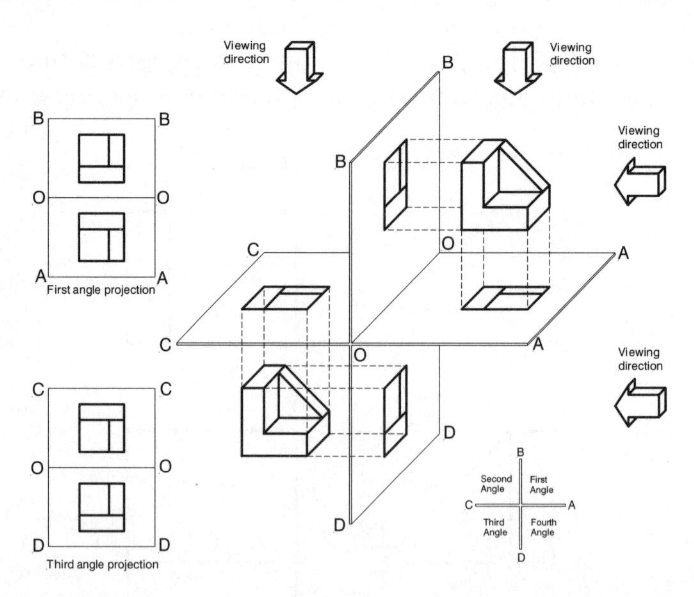

Figure 2.15 *Geometric construction showing the meaning of first and third angle projection*

180°, as shown in the small diagrams in Figure 2.15, the two views will be in first angle projection arrangement.

When the part in the third quadrant is viewed from the right-hand side and from above, the view will be projected forwards onto the faces OC and OD. When the planes are opened up to 180°, the views will be in third angle projection arrangement, as shown in the small diagrams in Figure 2.15.

If parts were to be placed in the second and fourth quadrant, the views projected onto the faces when opened out would be incoherent and invalid because they cannot be projected from one another. It is for this reason that there is no such thing as *second angle projection* or *fourth angle projection*.

There are several ISO standards dealing with views in first and third angle projection. These standards are: ISO 128:1982, ISO 128–30:2001 and ISO 128–34:2001.

2.7 Sectional views

There are some instances when parts have complex internal geometries and one needs to know information about the inside as well as the outside of the artefact. In such cases, it is possible to include a section as one of the orthographic views. A typical section is shown in Figure 2.16. This is a drawing of a cover that is secured to another part by five bolts. These five bolts pass through the five holes in the edge of the flange. There is an internal chamber and some form of pressurised system is connected to the cover by the central threaded hole. The engineering drawing in Figure 2.16 is in third angle projection. The top drawing is incomplete. It is only half the full flange. This is because the part is symmetrical on either side of the horizontal centre line, hence the 'equals' signs at either end. This means that, in the observer's eye, a mirror image of the part should be placed below the centre line. Note that the view projected (beneath) from this plan view is not a side view but a section through the centre. In museums, it is normal practice to cut or section complex parts like engines to show the internal workings. Parts that are sectioned are invariably painted red (or any other bright colour!). In engineering drawing terms, the equivalent of painting something red is to use cross-hatching lines which, in the case of Figure 2.16, are placed at 45°. The ISO rules concerning the form and layout of such section lines is given in Chapter 3. The method of indicating the fact that a section has been taken on the view, from which the section is projected, is shown in the plan view of the flange. Here, the centre line has two thicker lines at either end with arrows showing the direction of viewing. Against the arrows are the capital letters 'A', and it is along these lines and in the direction of arrows that the sectional view is taken. The third angle projection view beneath is a section along the line AA, hence it is given the title 'Section AA'. This method of showing the section position with a thickened line and arrows is explained further in the following chapter on ISO rules.

Other examples of sections are given in the assembly drawing of a small hand vice (see Figure 1.11) and the detailed drawing of the movable jaw of the vice (see Figure 1.12). In the case of the movable jaw detailed drawing in Figure 1.12, the front view is shown on the top-left and the right-hand side drawing view is a right-hand section through the centre line. In this instance there are no section lines or arrows to indicate that it is a section through the centre. However, in this case, it should be obvious that the section is through the centre and therefore it is not necessary to include the arrows. However, this is not the case for the inverted planned view, which is a complicated half-section with two section plane levels on the left-hand side and a

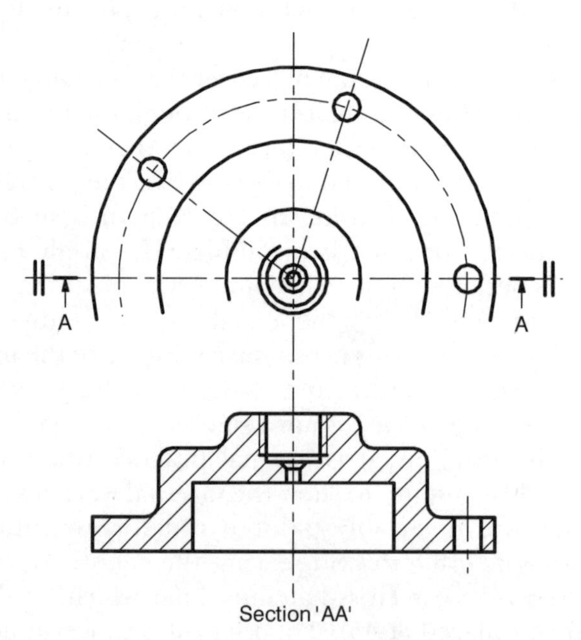

Figure 2.16 *Example of a sectional view of a flange*

conventional inverted plan (unsectioned) view on the right-hand side. Because this is a complicated inverted plan view, the section line and arrows are shown to guide the viewer. Note that the cross-hatched lines on the two different left-hand planes are staggered slightly.

A different type of section is shown in the assembly drawing in Figure 1.11. Here the movable jaw (part number 3), the hardened insert (part number 2), the bush (part number 4), the bush screw (part number 5) and part of the jaw clamp screw (part number 6) are shown in section. This is what is termed a 'local' section because the whole side view is not in section but a part of it. The various parts in the section are cross-hatched with lines at different slopes and different spacings. The section limits are shown by the zig-zag line on the movable jaw and a wavy line on the jaw clamp screw. Another type of section is shown on the tommy bar of the assembly drawing. This is a small circle with cross-hatching inside. This is called a 'revolved section' and it shows that, at this particular point along the tommy bar, the cross-sectional shape is circular. In this instance the cross-sectional shape would be the same at any point along the tommy so it doesn't really matter where the section appears.

The ISO standards dealing with sectional views are ISO 128–40:2001 and ISO 128–44:2001.

2.8 Number of views

In the examples of the cornflake packet shown in Figure 2.11 and the small bracket shown in Figure 2.12, six views of each component were shown. There can only ever be six views of an artefact in a full orthographic projection. The central view is invariably the front view.

Other views can be included but these will be auxiliary views. Such auxiliary views are placed remote from the orthographic views. If an artefact contains a sloping surface, the true view of the inclined surface will never be seen in orthographic projection. This can be seen in the small bracket in Figure 2.12. The bracket contains a stiffening wall which is shown on the right-hand side of the front view. This has a sloping surface as shown by the left-side view and the right-side view. However, there is no view that shows the true view of this place. This could be provided by an auxiliary view, projected from the left-side view or the right-side view that would be a view perpendicular to the inclined face. Such an inclined view would not fit comfortably within the six views of the bracket and therefore would be placed off at the side but with a note making clear that it was a view on an arrow perpendicular to the face. It is normal practice to label such arrows with some alphanumeric designation. There needs to be a title associated with the true view that relates the arrow to the view. A typical title would be 'View on arrow Z'.

Unless a part is very complex, six views of an artefact are unnecessary and over the top. The number of views will be dependent on the transmission of full and complete information of the artefact. Thus, considering the bracket in Figure 2.12, only three views would probably be needed. These would be the front view, the plan view and the left-side view. These three views would then be dimensioned and the three views plus the dimensions would be sufficient for the bracket to be made. Three such views are shown in Figure 2.14 (but the projection lines need to be rubbed out). Figure 2.16 shows two views of a flange. Since one view is a sectional view through the centre line, sufficient information can be transmitted when this part is dimensioned for it to be manufactured. In the small hand vice assembly drawing in Figure 1.11, three views are shown. The only reason that the left-hand view is shown is to give details of the screws (part number 8) which hold one of the hardened inserts to the body. An alternative method of drawing these bolts would be by adding dotted lines to the side view such that the hidden detail of the bolts was shown. In this case the balloon references would go to these dotted lines and the left-hand view would be unnecessary. However, I drew the three views because I thought it would be clearer than adding dotted lines.

The three drawings in Figure 1.11 are sufficient to assemble the various parts of the small hand vice. However, this cannot be said for the parts necessary to assemble George Stephenson's Rocket. In this case the two views would only give the barest of information about the outside shape and form. Numerous other views and indeed additional drawings would be needed to give full details on how to make and assemble the locomotive.

No hard and fast rules can be given with respect to the number of views required on any engineering drawing. The decision on the number required will be dependent on the complexity of the artefact and its internal features. In all cases the number of views will be driven by the need to give sufficient information for the part to be manufactured. One should try to avoid giving more views than is necessary because this just tends to complicate a drawing. On the other hand, if an extra view helps in the understanding of the part design, then it is a useful addition!

References and further reading

Bailey R B and Glithero J P, *The Engineering History of the Rocket, a Survey Report*, National Railway Museum, York, 2000.

ISO 128:1982, *Technical Drawings – General Principles of Presentation*, 1982.

ISO 128–24:1999, *Technical Drawings – General Principles of Presentation – Part 24: Lines on Mechanical Engineering Drawings*, 1999.

ISO 128–30:2001, *Technical Drawings – General Principles of Presentation – Part 30: Basic Conventions for Views*, 2001.

ISO 128–34:2001, *Technical Drawings – General Principles of Presentation – Part 34: Views on Mechanical Engineering Drawings*, 2001.

ISO 128–40:2001, *Technical Drawings – General Principles of Presentation – Part 40: Basic Conventions for Cuts and Sections*, 2001.

ISO 128–44:2001, *Technical Drawings – General Principles of Presentation – Part 44: Sections on Mechanical Engineering Drawings*, 2001.

ISO 5456–1:1996, *Technical Drawings – Projection Methods – Part 1: Synopsis*, 1996.

ISO 5456–2:1996, *Technical Drawings – Projection Methods – Part 2: Orthographic Representations*, 1996.

ISO 5456–3:1996, *Technical Drawings – Projection Methods – Part 3: Axonometric Representations*, 1996.

3

ISO Drawing Rules

3.0 Introduction

In the previous chapter, a comparison was made between engineering drawing and the rules of sentence construction and grammar of a written language. Chapter 2 was concerned with the definition of the artefact shape and form and the ISO rules which define how a 3D artefact is to be drawn on a 2D drawing sheet. In this chapter, information is given on how to specify the manufacturing requirements. It covers such things as size, shape, dimensions, tolerances, surface finish and assembly specifications.

3.1 Example of drawing a small hand vice

A common artefact in any workshop is a small vice. Such a small engineering vice is shown in Figure 3.1. The main body of the vice is a stubby 'U' shape in which a movable jaw is positioned between the two uprights. The movable jaw is actuated by a screw which is rotated by a small bar. Although the drawing is 'busy', the different lines help to make the artefact jump out from the page. This has been done by the use of different types of line thicknesses (thick and thin) and different types of line styles (continuous, discontinuous, dash, chain dotted). This is an assembly drawing and is not meant to provide any manufacturing details. The individual components making up the vice are numbered using a 'balloon' reference system, i.e., small circles with the part numbers in them. The assembly drawing is of little use on its own because it needs a list to identify each individual part within the assembly. Such a parts list or item list is shown as part of the drawing in Figure 3.1.

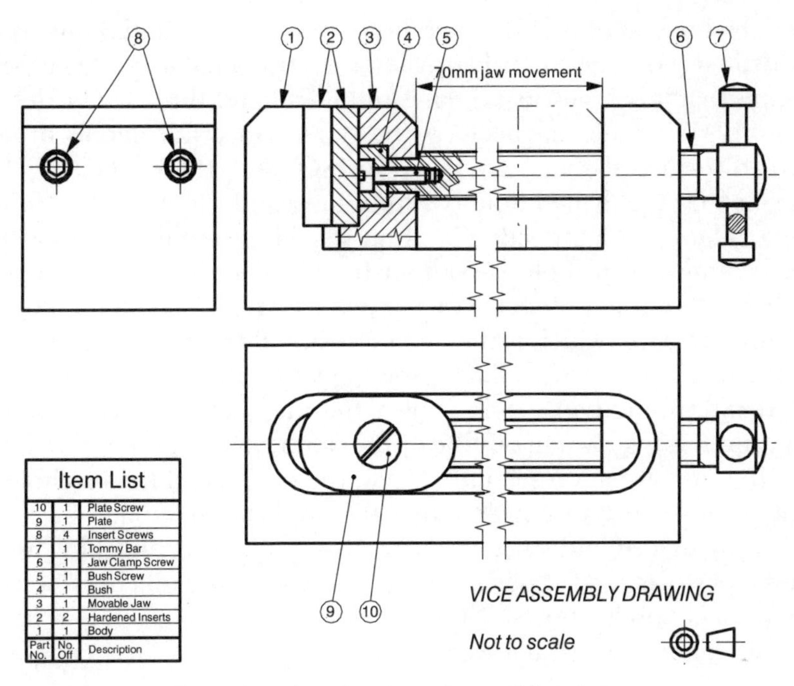

Figure 3.1 *Assembly engineering drawing of a small hand vice*

Figure 3.2 is a third-angle orthographic projection 'detail' drawing of the movable jaw (part number 3). It gives all the information necessary for the part to be manufactured. The outline is drawn in thick (or wide) lines whereas additional information (e.g. hidden detail or section hatching) is drawn in thin (or narrow) lines. The thick lines are deliberately drawn so that shape and form 'jump' out of the picture. With regard to the front elevation, the 'equals' sign at either end of the centre line shows that it is symmetrical about that centre line. The 16mm wide tongue is thus centrally positioned in the front elevation and there is no need to dimension its position from either side. There are further outcomes from this symmetry. Firstly, both underside surfaces that contact the body (as shown by thick chain dotted lines) are to be polished such that the average surface finish (Ra) is less than 0,2um. Secondly, the counter-bored 5mm diameter holes are identical. The right-hand elevation is a section through the centre of the jaw but nothing tells you this. This is the designer's decision of how much to include in the drawing, called 'draughtsman's licence'. The side elevation shows that there is a vertical threaded hole in the base. The various line thicknesses of the threaded hole show that the initial hole is to be drilled (note the conical end) and then threaded to M8. The 'M8' means that it is a metric standard 8mm diameter thread. The designation 'M8' is all that needs to be stated since full details of the thread form and shape are given in ISO 68–1:1998. The 'x10/12' means that the drilled hole is 12mm long and the thread is 10mm long. The right-hand side elevation section also indicates that the horizontal central hole is counter-bored. The dimensions of this hole are shown in note form on the inverted plan. The initial hole is 10mm diameter which is then counter-bored to 15mm diameter to a depth of 7.5mm with a flat bottom (given by the 'U'). The position of the hardened insert is shown on the sectioned right-hand elevation. It is shown in outline by the double chain dotted thin line. On the side elevation sectioned view, the position of the M8 hole is not given. In such instances as this, the implication is that the hole is centrally placed and since its exact position is not critical for functional performance, it perhaps does not matter too much. However, in product liability terms, all dimensions should be given and none left to chance. Thus, if I were drawing this for real in a company

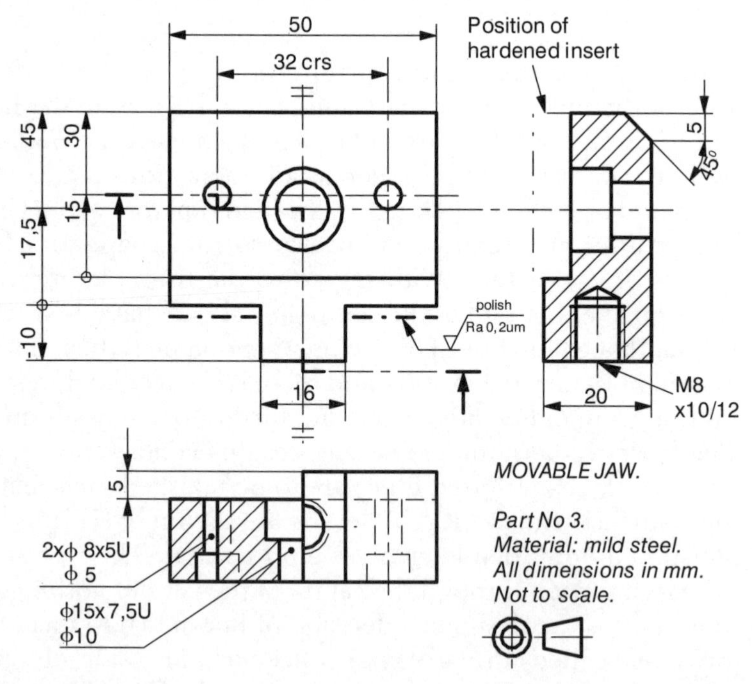

Figure 3.2 *Detailed engineering drawing of the 'movable jaw', part number 3*

I would label its position as 10mm from the left-hand or the right-hand side. However, to illustrate the point, I have left it off the drawing. The inverted plan (lower left-hand drawing) is a staggered section projected from the front elevation. The staggered section lines are shown by the dual thick and thin chain dotted lines terminating in arrows that give the direction of viewing. Thus, the inverted plan is a part section.

Figure 3.3 shows a detail drawing of the hardened insert (part number 2). This illustrates some other principles and applications of engineering drawing practice. Two views are shown. Note that the hardened insert is symmetrical as shown by the centre line and the 'equals' symbols at each end. Hence, I chose only to show one half. With regard to the left-hand side elevation, the side is flame hardened to provide abrasion resistance. The 'HRC' refers to the Rockwell 'C' hardness scale. The M5 threaded hole is 15mm from the lower datum place and the hole insert is 30mm high. The M5 hole could have been shown as being symmetrical with 'equals' signs on the other centre line instead of being dimensioned from the base.

Only two detail drawings (Figures 3.2 and 3.3) are shown for convenience. If this were a real artefact that really was to be manufactured, detailed drawings would be required for all the other parts. However, there is no need to provide detailed drawings of standard items like the screws.

3.2 Line types and thicknesses

The standard ISO 128:1982 gives 10 line types that are defined A to K (excluding the letter I). The table in Figure 3.4 shows these lines.

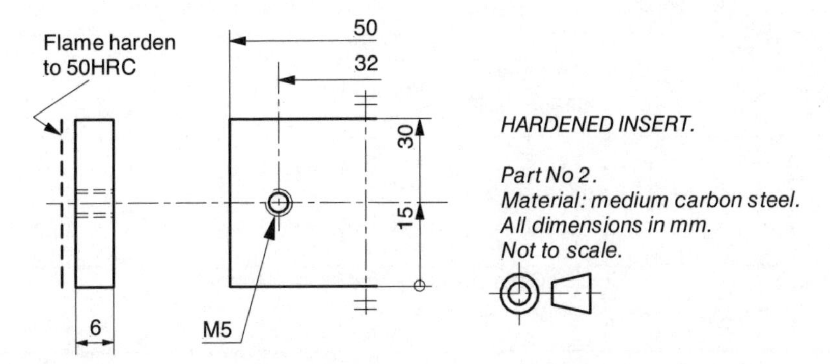

Figure 3.3 *Detailed engineering drawing of the 'hardened insert', part number 2*

The line types are 'thick', 'thin', 'continuous', 'straight', 'curved', 'zigzag', 'discontinuous dotted' and 'discontinuous chain dotted'. Each line type has clear meanings on the drawing and mixing up one type with another type is the equivalent of spelling something incorrectly in an essay.

The line thickness categories 'thick' and 'thin' (sometimes called 'wide' and 'narrow') should be in the proportion 1:2. However, although the proportion needs to apply in all cases, the individual line thicknesses will vary depending upon the type, size and scale of the drawing used. The standard ISO 128:1982 states that the thickness of the 'thick' or 'wide' line should be chosen according to the size and type of the drawing from the following range: 0,18; 0,25; 0,35; 0,5; 0,7; 1; 1,4 and 2mm. However, in a direct contradiction of this the standard ISO 128–24:1999 states that the thicknesses should be 0,25; 0,35; 0,5; 0,7; 1; 1,4 and 2mm. Thus confusion reigns and the reader needs to beware! With reference to the table in Figure 3.4, the A–K line types are as follows.

The ISO type 'A' lines are thick, straight and continuous, as shown in Figure 3.5. They are used for visible edges, visible outlines, crests of screw threads, limit of length of full thread and section viewing lines. The examples of all these can be seen in the vice assembly detailed drawings. These are by far the most common of the lines types since they define the artefact.

The ISO type 'B' lines are thin, straight and continuous, as shown in Figure 3.6. They are used for dimension and extension lines,

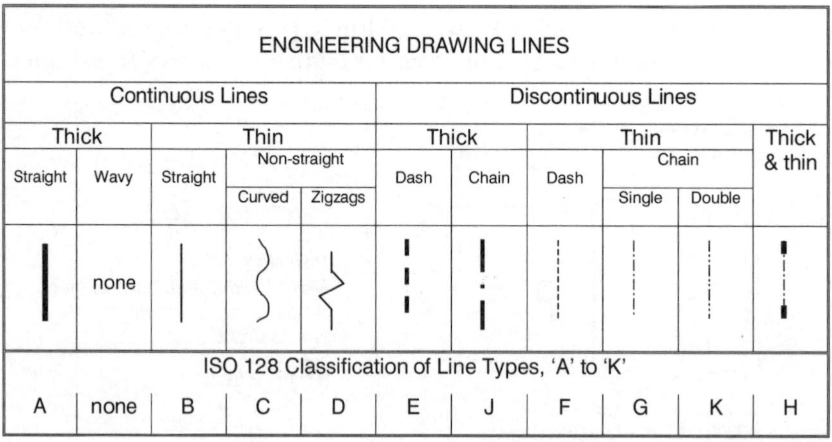

ENGINEERING DRAWING LINES										
Continuous Lines					Discontinuous Lines					
Thick		Thin			Thick		Thin			Thick & thin
			Non-straight					Chain		
Straight	Wavy	Straight	Curved	Zigzags	Dash	Chain	Dash	Single	Double	
	none									
ISO 128 Classification of Line Types, 'A' to 'K'										
A	none	B	C	D	E	J	F	G	K	H

Figure 3.4 *Engineering drawing line types A to K (ISO 128:1982)*

leader lines, cross hatching, outlines of revolved sections, short centre lines, thread routes and symmetry ('equals') signs.

The ISO type 'C' lines are thin, wavy and continuous, as shown in Figure 3.7. They are only used for showing the limits of sections or the limits of interrupted views as would be produced by freehand drawings by a draughtsman on a paper-based drawing board. Examples of type 'C' lines are shown on the assembly drawing, part number six, jaw clamp screw.

The ISO type 'D' lines are thin, zigzag and continuous, as shown in Figure 3.8. These have exactly the same use as the type 'C' lines

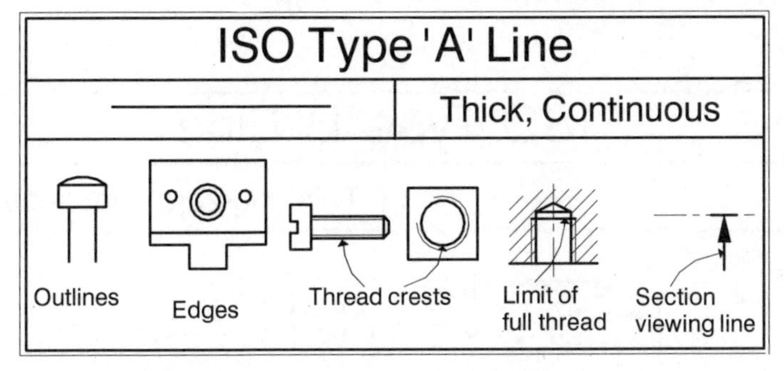

Figure 3.5 *ISO 128 engineering drawing line type 'A'*

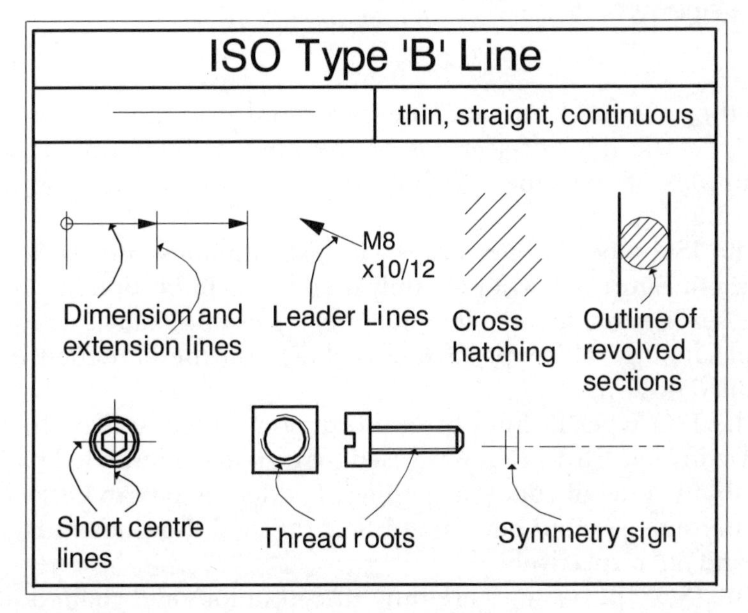

Figure 3.6 *ISO 128 engineering drawing line type 'B'*

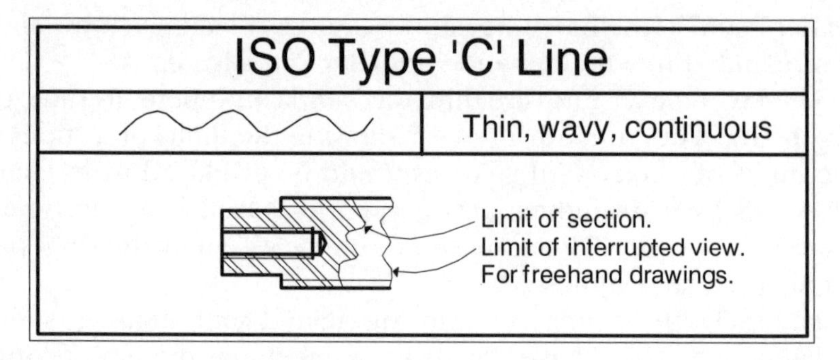

Figure 3.7 *ISO 128 engineering drawing line type 'C'*

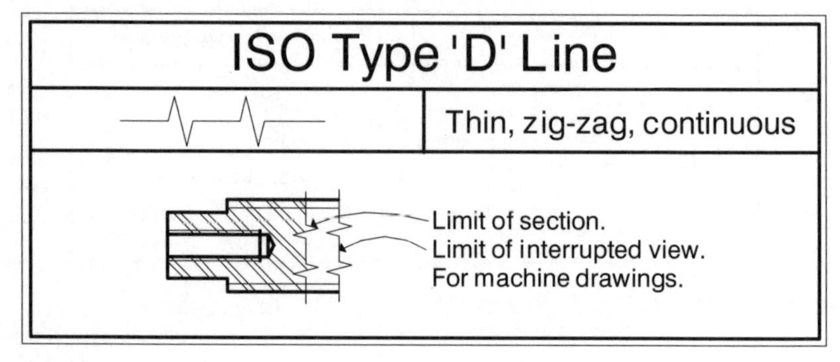

Figure 3.8 *ISO 128 engineering drawing line type 'D'*

but they are used for machine-generated drawings. Again they apply to the limit of sections or the limit of interrupted views. Examples of the type 'D' line are shown in the vice assembly drawing.

The ISO type 'E' lines are thick, discontinuous and dashed, as shown in Figure 3.9. They are only used for an indication of permissible surface treatment. This could be, for example, heat treatment or machining. This type of line is shown on the hardened insert detailed drawing.

The ISO type 'F' lines are thin, discontinuous and dashed, as shown in Figure 3.10. They are used for displaying hidden detail, be that hidden detail edges or outlines. Hidden detail can be seen on the movable jaw and hardened insert detailed drawings in Figures 3.2 and 3.3 respectively.

The ISO type 'G' lines are thin, discontinuous and chain dotted, as shown in Figure 3.11. They are used to show centre lines of either

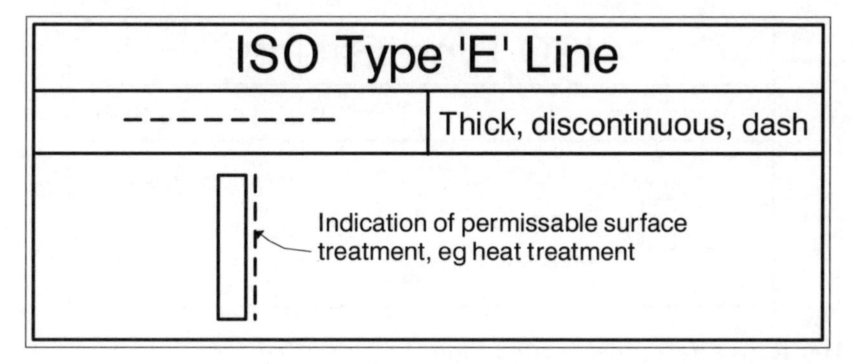

Figure 3.9 *ISO 128 engineering drawing line type 'E'*

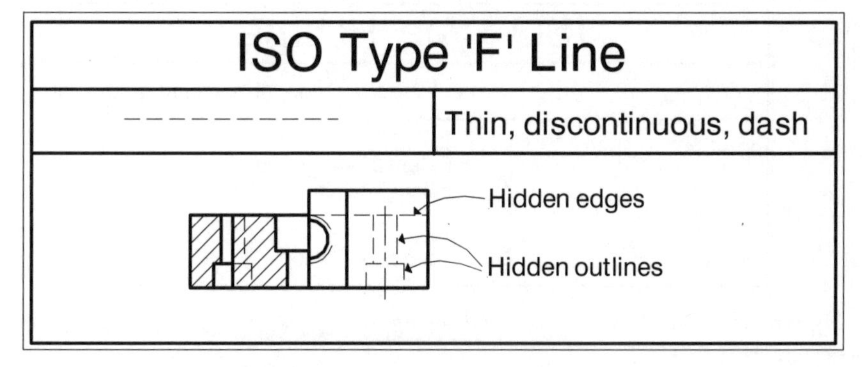

Figure 3.10 *ISO 128 engineering drawing line type 'F'*

individual features or parts. Centre lines can be seen on the vice assembly drawing as well as the movable jaw and hardened insert drawings.

The ISO type 'H' lines are a combination of thick and thin, discontinuous and chain dotted, as shown in Figure 3.12. They are used to show cutting planes. The thick part of the type lines are at the ends where the cutting section plain viewing direction arrows are shown as well as at the points of a change in direction. An example of a staggered type 'H' cutting plane is shown in the movable jaw detailed drawing.

Note that no line type 'I' is defined in the ISO 128:1982 standard.

The ISO type 'J' lines are thick, discontinuous and chain dotted, as shown in Figure 3.13. They are used for the end parts of cutting planes as shown previously in the above type 'H' lines. They are also used to provide an indication of areas that are limited for some

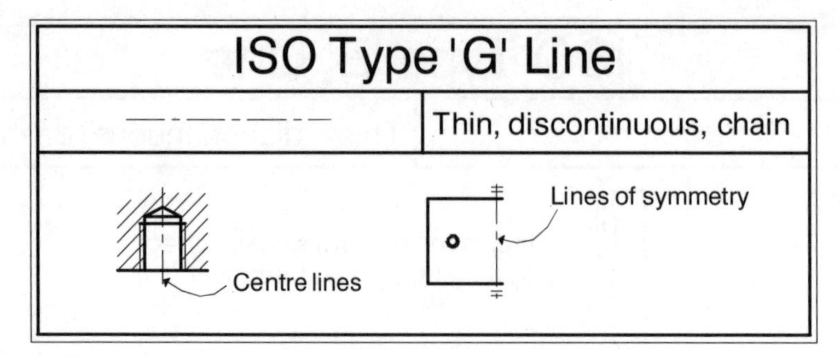

Figure 3.11 *ISO 128 engineering drawing line type 'G'*

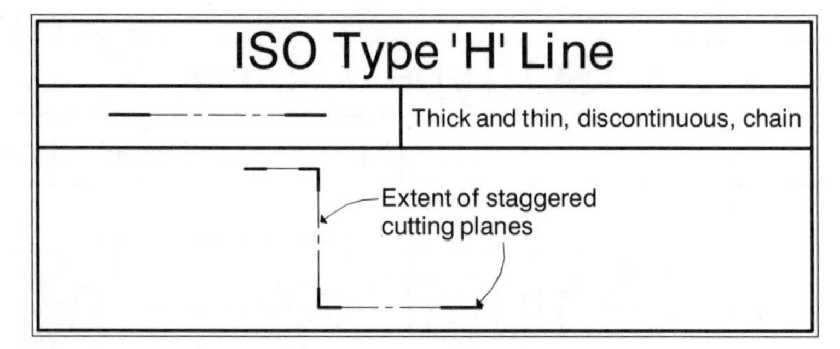

Figure 3.12 *ISO 128 engineering drawing line type 'H'*

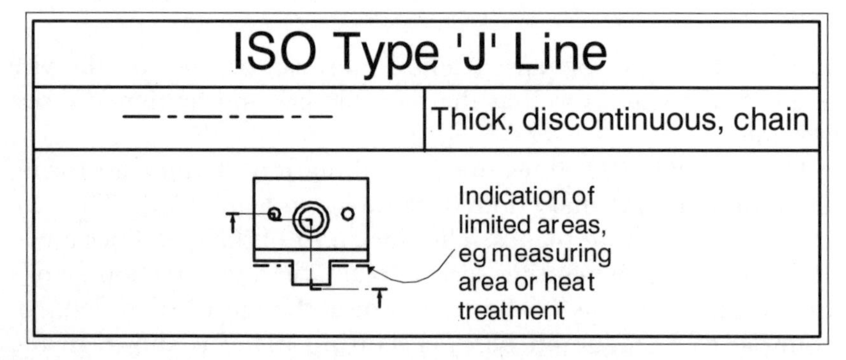

Figure 3.13 *ISO 128 engineering drawing line type 'J'*

reason, e.g. a measuring area or a limit of heat-treatment. Examples of this type of line can be seen in the movable jaw detailed drawing.

The ISO type 'K' lines are thin, discontinuous and chain dotted with a double dot, as shown in Figure 3.14. They are used to indicate the important features of other parts. This could be either the

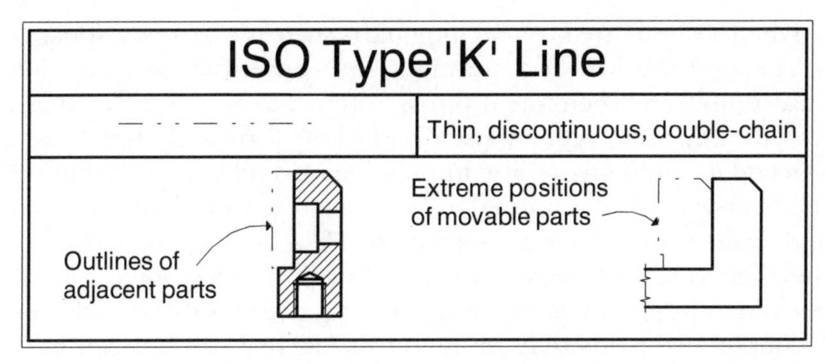

Figure 3.14 *ISO 128 engineering drawing line type 'K'*

outline of adjacent parts to show where a particular part is situated, or, for movable parts, the extreme position of movable parts.

3.3 Sectioning or cross-hatching lines

When you go to a museum, you often see artefacts that have been cut up. For example, to illustrate how a petrol engine works, the cylinder block can be cut in half and the cut faces are invariably painted red. In engineering drawing, cross-hatching is the equivalent of painting something red. It is used to show the internal details of parts which otherwise would become too complex to show or dimension.

The cross-hatch lines are usually equi-spaced and, for small parts, cover the whole of the 'red' cut area. They are normally positioned at 45° but if this is awkward because the part itself or a surface of it is at 45°, the hatching lines can be at another angle. Logical angles like 0°, 30°, 60° or 90° are to be preferred to peculiar ones like 18° (say). If sectioned parts are adjacent to each other, it is normal to cross hatch in different orientations (+ and –45°) or if the same orientation is used, to use double lines or to stagger the lines. Examples of single and double + and –45° cross-hatching lines are shown in the vice assembly drawing in Figure 3.1. An example of staggered cross-hatching is shown in the inverted plan drawing of the movable jaw in Figure 3.2.

If large areas are to be sectioned, there is no particular need to have the cross-hatching lines covering the whole of the component but rather the outside regions and those regions which contain details.

When sections are taken of long parts such as ribs, webs, spokes of wheels and the like, it is normally the convention to leave them unsectioned and therefore no cross-hatch lines are used. The reason for this is that the section is usually of a long form such that if it were hatched it would give a false impression of rigidity and strength. In the same way it is not normal to cross hatch parts like nuts and bolts and washers when they are sectioned. These are normally shown in their full view form unless, for example, a bolt has some specially machined internal features such that it is not an off-the-shelf item. Example of threads that are not cross-hatched can be seen in the vice assembly drawing in Figure 3.1.

3.4 Leader lines

A leader line is a line referring to some form of feature that could be a dimension, an object or an outline. A leader line consists of two parts. These are:

- A type B line (thin, continuous, straight) going from the instruction to the feature.
- A terminator. This can be a dot if the line ends within the outline of the part, an arrow if the line touches the outline or centre line of a feature or without either an arrowhead or a dot if the line touches a dimension.

Examples of leader lines with arrowheads and dots are shown in the vice assembly and the movable jaw drawings.

3.5 Dimension lines

Various ISO standards are concerned with dimensioning. They are under the heading of the ISO 129 series. The basic standard is ISO 129:1985 but it has various parts to it.

A dimensioning 'instruction' must consist of at least four things. Considering the 50mm width of the jaw and the 32mm spacing of the holes of the movable jaw drawing in Figure 3.15, these are:

- Two projection lines which extend from the part and show the beginning and end of the actual dimension. They are projected from the part drawing and show the dimension limits. In Figure

3.15, the width is 50mm and the projection lines for this dimension show the width of the part. They are type B lines (thin, continuous and straight). These lines touch the outline of the part. The projection lines for the hole-centre spacing dimension of 32mm are centre lines. They are type G lines (thin, discontinuous, chain) which pass through the drawing just past where the holes are located.

- A dimension line which is a type B line (thin, continuous and straight). In Figure 3.15, these dimension lines are the length of the dimension itself, i.e. '50' or '32' mm long.
- A numerical value which is a length or an angle. In the Figure 3.15 example the dimensions are the '50mm' and '32mm' values. If a part is not drawn full size because it is too small or too large with respect to the drawing sheet, the actual dimension will be the value which it is in real life whereas the dimension line is scaled to the length on the drawing.
- Two terminators to indicate the beginning and end of the dimension line. The terminators of '50' and '32' dimensions in Figure 3.15 are solid, narrow arrowheads. Other arrowhead types may be used. There are four types of arrowhead allowed in ISO, as shown in Figure 3.16. These four are the narrow/open (15°), the wide/open (90°), the narrow/closed (15°) and the narrow/solid (15°). An alternative to an arrowhead is the oblique stroke. When several dimensions are to be projected from the same position, the 'origin' indication is used, consisting of a small circle. These drawings are shown in Figure 3.16. An example of an origin indicator is shown in the movable jaw detailed drawing.

Many dimensioning examples can be seen in the movable jaw and hardened insert detail drawings. The dimensions in these two drawings follow the following convention. All terminators are of the solid arrow type, all projection lines touch the outside of the part outline, all dimension numerical values are placed above the dimension lines and all dimension values can be read from the left-hand bottom corner of the drawings.

The dimensioning convention used in the movable jaw and hardened insert detail drawings is the one which is the most commonly used one. However, alternative dimensioning conventions are allowed in the ISO standards. These will be covered in Chapter 4.

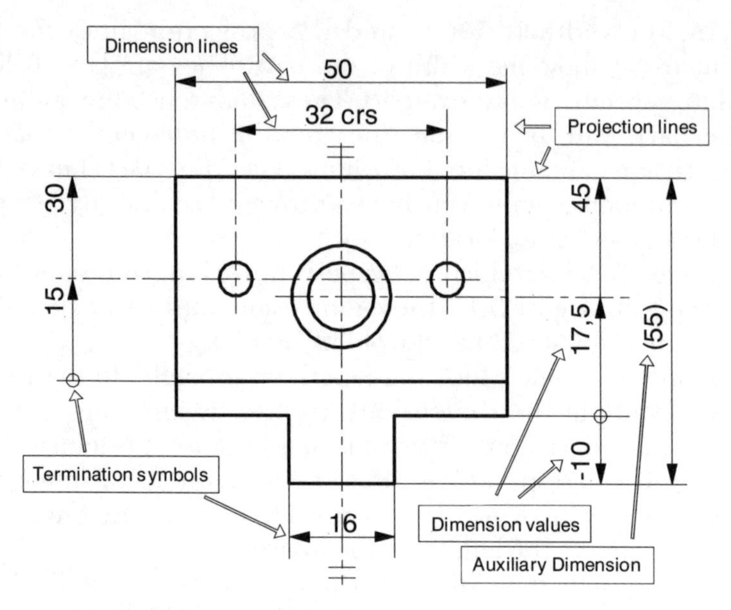

Figure 3.15 *Example of general dimensioning*

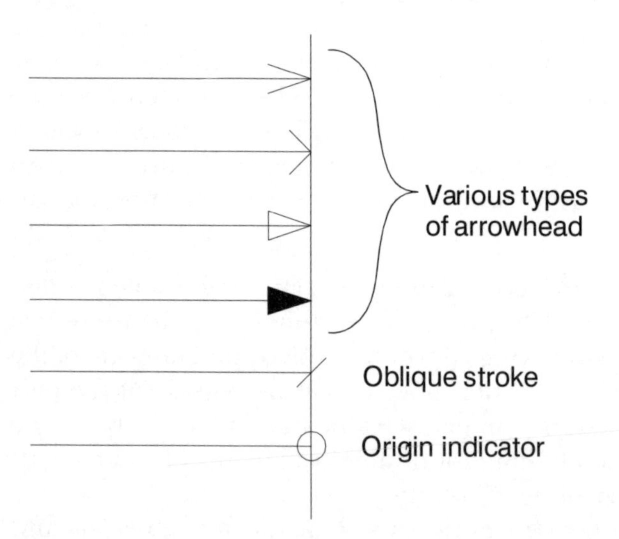

Figure 3.16 *The various types of dimension line terminators*

3.6 The decimal marker

Readers in the UK and USA should be aware that a full stop or point is no longer recommended as the decimal marker. The ISO recommended decimal marker is now the comma. Thus, taking pi as an example, it should now be written as 3,142 and not 3.142. Similarly, the practice of using a comma as a 10^3 separator is no longer recommended. A space should be used instead. Thus, one million should be represented as 1 000 000 and not 1,000,000.

3.7 Lettering, symbols and abbreviations

Many drawings are microfilmed and this causes a problem of legibility when drawings are blown up again to their original size. Thus, it is recommended that the distance between adjacent lines or the spacing between letters or numerals should be at least twice the line thickness. There are six ISO standards (would you believe it?) on lettering alone; they are under one standard. The six parts of ISO 3098 refer to: general requirements (part 0), the Latin alphabet (part 2), the Greek alphabet (part 3), diacritical marks (part 4), CAD lettering (part 5) and the Cyrillic alphabet (part 6).

Symbols and abbreviations are used on engineering drawings to save space and time. However, because they are shorthand methods they need to impart precise and clear information. Standard English language symbols and abbreviations are shown in the BSI standard BS 8888:2000. Various abbreviations can be seen in the movable jaw detailed drawing in Figure 3.2. The 'CRS' refers to the fact that the hole centre lines are 32mm apart and the Greek letter 'ϕ' is used to indicate diameter. Other symbols and abbreviations are covered in Chapter 4.

Standard screw thread and threaded part dimensions are detailed in ISO 68–1 and ISO 6410, parts 1, 2 and 3:1993. Thus, the only symbol which needs to appear with respect to a threaded part is the 'M' of the threaded hole on the right-hand end elevation section drawing. There are other abbreviations concerned with holes that are not covered by the BS 8888:2000 standard. These are the abbreviations and symbols and shorthand methods associated with the dimensioning of holes, whether they are plain, threaded or stepped. For example, the M8 threaded hole has the numbers '10' and '12' separated by a forward slash. This means that the drilled hole is 12mm deep and the threaded section 10mm long. The notes referring to the countersunk holes on the inverted plan sections use the abbreviation 'U'. This refers to a flat-bottomed hole whether it be a counter-bored or a full hole. If a hole were required to have a vee-shaped hole bottom, the symbol 'V' would be used. There is a complete standard concerned with the symbology and the abbreviations associated with holes; this is ISO 15786:2001. This hole symbology is considered again in Chapter 4.

3.8 Representation of common parts and features

There are several standard feature shapes and forms that can be represented in a simplified form, so saving drawing time and cost. The most common types are covered below.

3.8.1 Adjacent parts

In a detailed drawing of a particular part, it may be necessary to show the position of adjacent part/s for the convenience of understanding the layout. In the case of the detailed drawing of the movable jaw, the adjacent hardened jaw position is indicated by the chain double-dotted line on the left-hand side of the right-hand-side sectional view. Such parts need to stand out but not be obtrusive so they are drawn using type K lines, the thin, continuous, double chain dotted lines. Adjacent parts are usually shown in outline without any specific details.

3.8.2 Flats on cylindrical or shaped surfaces

It is not always obvious that surfaces are flat when they are on otherwise curved, cylindrical or spherical surfaces. In this case, flat surfaces such as squares, tapered squares and other flat surfaces may be indicated by thin 'St Andrew' cross type diagonal lines. An example of this is shown in the entirely fictitious gear shaft in Figure 3.17. The extreme right-hand end of the shaft has a reduced diameter and approximately half of this cylindrical length has been flat milled to produce a square cross-section. The fact that the cross-sectional shape of this region is square and not cylindrical is seen in the end view as a square and in the right-hand side elevation by the crosses.

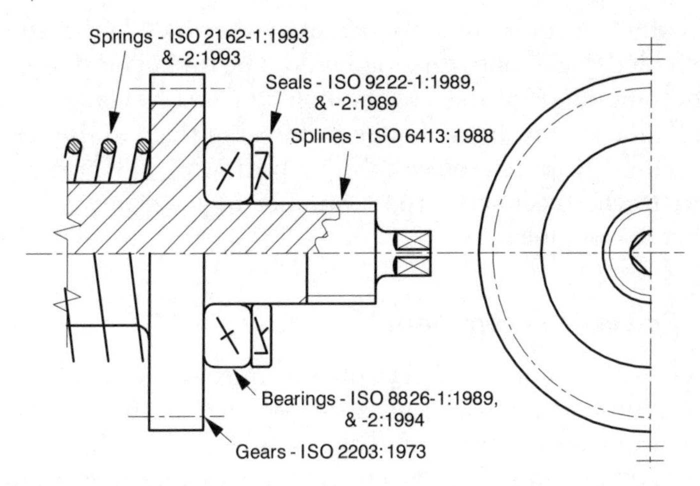

Figure 3.17 *A fictitious gear shaft with bearings, seals, springs and splines with the relevant ISO references*

3.8.3 Screw threads

Screw threads are complex helical forms and their detailed characteristics in terms of such things as angles, root diameter, pitch circle diameter and radii are closely defined by ISO standards. Thus, if the designation 'M8' appears on a drawing it would appear at first sight to be very loosely defined but this is far from the case. Screw threads are closely defined in the standard ISO 6410, parts 1, 2 and 3:1993. The 'M8' designation automatically refers to the ISO 68–1:1998, ISO 6410–1, 2 and 3:1993 standards in which things like the thread helix angle, the vee angles and the critical diameters are fully defined. Thus, as far as screw threads are concerned, there is no need to do a full drawing of a screw thread to show that it is a screw thread. This takes time and costs money. The convention for drawing an engineering thread is shown using a combination of ISO type A and B lines as shown in the drawings in Figures 3.1, 3.2 and 3.3. A screw thread is represented by two sets of lines, one referring to the crest of the thread (type A line) and the other referring to the roots of the thread (type B line). These can be seen for a bolt and a hole in Figures 3.5 and 3.6. This representation can be used irrespective of the exact screw thread. For example, on the vice assembly drawing in Figure 3.1, the screw thread on the bush screw (part number 5) and the jaw clamp screw (part number 6) are very different. In the real vice, the former is a standard vee-type thread whereas the latter is a square thread.

Line thicknesses become complicated when a male-threaded bolt is assembled in a female-threaded hole. The thread crest lines of the bolt become the root lines of the hole and vice versa. This means that in an assembly, lines change from being thick to thin and vice versa. This is shown in the vice assembly drawing in Figure 3.1, with respect to the bush screw (part number 5)/jaw clamp screw (part number 6) assembly.

3.8.4 Splines and serrations

Splines and serrations are repetitive features comparable to screw threads. Similarly, it is not necessary to give all the details of the splines or serrations, the symbology does it for you. The convention is that one line represents the crests of the serrations or splines and the other the roots. This is shown in the hypothetical drawing in Figure 3.17 where there is a spline at the right-hand end of the gear drive shaft. A note would give details of the spline. The standard ISO 6413:1988 gives details of the conventions for splines.

3.8.5 Gears

Gear teeth are a repetitive feature similar to screw threads or splines. It is not necessary to show their full form. In non-sectional views, gears are represented by a solid outline without teeth and with the addition of the pitch diameter surface of a type G line. In a transverse section, the gear teeth are unsectioned whereas the body of the gear is. The limit of the section hatching is the base line of the teeth as shown in the drawing in Figure 3.17. In an axial section, it is normal to show two individual gear teeth unsectioned but at diametrically opposed positions in the plane of the section. All details of the gear type shape and form need to be given via a note. In a gear assembly drawing which shows at least two gears, the same principle as for individual teeth (above) is used but at the point of mesh, neither of the two gears is assumed to be hidden by the other in a side view. Both of the gears' outer diameters are shown as solid lines. The standard ISO 2203:1973 gives details of the conventions for gears.

3.8.6 Springs

It is not normal to show the full shape and form of springs. Their helical form means very complicated drawing shapes. The simplified representation is a zig-zag shape of ISO type A lines for side views. If the spring is shown in cross-section, the full form is drawn as is shown in Figure 3.17. A note should provide all the fine details of the spring design. The standards ISO 2162–1:1993, ISO 2162–2:1993 and ISO 2162–3:1993 give details of the conventions for springs.

3.8.7 Bearings

As shown in the simplified gear shaft assembly in Figure 3.17, the transverse view of a bearing is shown in cross-section with only the outline and none of the internal details such as ball bearings and cages. Even when the transverse view is not a sectional view, it is normal practice to show the bearing as if it was a cross-sectional view. Within the bearing outline (class A line), symbology is used to indicate the exact type of bearing. Symbology shown within the example in Figure 3.17 refers to a thrust bearing. Details of how to draw bearings are covered in ISO 8826–1:1989 and ISO 8826–2:1994.

3.8.8 Seals

Seals are treated in almost in the same way as gears. This is shown in the shaft assembly in Figure 3.17 where a seal is shown adjacent to the bearing. The outline of the seal is given and symbology within the outline shows the type of seal. In this case the seal is a lip type seal with a dust lip. The standards ISO 9222–1:1989 and ISO 9222–2:1989 give details of the various types of seal.

3.9 Item references and lists

In an assembly drawing, the various components or items which make up the assembly need to be referenced. In the vice assembly drawing in Figure 3.1, the individual items are shown by the 'balloon' reference system using the numbers 1–10 (in this case). For convenience the balloon item references are normally arranged in horizontal or vertical alignments. The small circles surrounding each number are optional. The standard that gives details of item references is ISO 6433:1981.

The list of items appropriate to the assembly drawing of the vice is shown in Figure 3.1. In this case three columns are shown, the item reference (part number) the number of each component required in the assembly (number of) and the description of the item. In this case there are three columns of information. Other columns can be added as appropriate. Examples of other columns are: material, stock number, delivery date, remarks and relevant ISO standards.

The vertical sequence of the item entries should be in numerical order. When the item list is included on the assembly drawing as in Figure 3.1, the items should read from bottom to top in numerical order with the column headings at the very bottom. However, if the item list is on a separate drawing, on its own, the sequence is to be from top to bottom with the headings at the top. The standard that gives details of item lists is ISO 7573:1983.

3.10 Colours

Colours are not normally used in engineering drawing. Indeed, in ISO 128:1982, the use of colour is 'not recommended'. The reason for this is for the convenience of document transmission that can be more easily achieved if the colour is always black. Hence, the standards recommend the use of the different line thicknesses and line designations such that discrimination is obtained without the use of colour.

3.11 Draughtman's licence

The term *draughtsman's licence* refers to the freedom a draughtsman has in expressing the design in drawing form. This applies, irrespective of whether a drawing is drawn by hand or on a CAD system. Any component can be represented in a variety of ways in terms of the drawing convention (i.e., number of views, sections, viewing direction, etc.) and the method of imparting the manufacturing details (i.e., the dimensions, tolerances and datum surfaces). The problem is that there are as many ways of drawing a part as there are draughtsmen or indeed draughtswomen. For example, in Figure 3.3, for the detailed drawing of the hardened insert I chose to only draw the left-hand part of the front elevation. I could just as well have drawn the whole front elevation but by doing it this way, I have saved myself time and money. The choice was mine, and in this case, I decided that the short cut method would not overly confuse anyone who read the drawing.

Other examples are shown in the drawings of the movable jaw in Figure 3.2. This is a complex part with many holes and in addition it has steps and a chamfer. The added complication is that the holes are slightly offset. Thus, I chose to draw a full front elevation with a sectioned right-hand side elevation and a part-sectioned inverted plan. The inverted plan incorporates three separate 'views'. The left-hand part has two sectional planes whereas the right-hand part of the inverted plan is unsectioned. This has allowed me to incorporate a variety of information on the one inverted plan.

The golden rule is that the designer should always avoid ambiguity and include as much information as possible to ensure that the part is returned from a subcontractor without any queries. There are two dangers in the transmission of information between the designer and the manufacturer. Firstly, information may be missing or may be ambiguous such that emails or faxes need to be passed backwards and forwards to clarify the situation. This costs extra time and money! Secondly, the last thing the draughtsman wants to have is a subcontracted part not assembling with all the other parts in the assembly because the subcontractor has interpreted the drawing in a manner that the designer did not intend. The draughtsman should always take an 'upper bound' approach when deciding how far he should go with his draughtsman's licence to minimise the influence of errors and ambiguities.

References and further reading

BS 8888:2000, *Technical Product Documentation – Specification for Defining, Specifying and Graphically Representing Products*, 2000.

ISO 68–1:1998, *General Purpose Screw Threads – Basic Profile Part 1: Metric Screw Threads*, 1998.

ISO 128:1982, *Technical Drawings – General Principles of Presentation*, 1982.

ISO 128–24:1999, *Technical Drawings – General Principles of Presentation – Part 24: Lines on Mechanical Engineering Drawings*, 1999.

ISO 129:1985, *Technical Drawings – Dimensioning – General Principles, Definitions, Methods of Execution and Special Indications*, 1985.

ISO 129–1.2:2001, *Technical Drawings – Indication of Dimensions and Tolerances – Part 1: General Principles*, 2001.

ISO 2162–1:1993, *Technical Product Documentation – Springs – Part 1: Simplified Representation*, 1993.
ISO 2162–2:1993, *Technical Product Documentation – Springs – Part 2: Presentation of Data for Cylindrical Helical Compression Springs*, 1993.
ISO 2163–3:1993, *Technical Product Documentation – Springs – Part 3: Vocabulary*, 1993.
ISO 2203:1973, *Technical Drawings – Conventional Representation of Gears*, 1973.
ISO 3098–0:1998, *Technical Drawings – Lettering – Part 0: General Requirements*, 1998.
ISO 3098–2:2000, *Technical Product Documentation – Lettering – Part 2: Latin Alphabet, Numerals and Marks*, 2000.
ISO 3098–3:2000, *Technical Product Documentation – Lettering – Part 3: Greek Alphabet*, 2000.
ISO 3098–4:2000, *Technical Product Documentation – Lettering – Part 4: Diacritical and Particular Marks for the Latin Alphabet*, 2000.
ISO 3098–5:2000, *Technical Product Documentation – Lettering – Part 5: CAD Lettering for the Latin Alphabet, Numerals and Marks*, 2000.
ISO 3098–6:2000, *Technical Product Documentation – Lettering – Part 6: Cyrillic Alphabet*, 2000.
ISO 6410–1:1993, *Technical Drawings – Screw Threads and Threaded Parts – Part 1: General Conventions*, 1993.
ISO 6410–2:1993, *Technical Drawings – Screw Threads and Threaded Parts – Part 1: Screw Threaded Inserts*, 1993.
ISO 6410–3:1993, *Technical Drawings – Screw Threads and Threaded Parts – Part 3: Simplified Representation*, 1993.
ISO 6413:1988, *Technical Drawings – Representation of Splines and Serrations*, 1998.
ISO 6433:1981, *Technical Drawings – Item References*, 1981.
ISO 7573:1983, *Technical Drawings – Item Lists*, 1983.
ISO 8826–1:1989, *Technical Drawings – Roller Bearings – Part 1: General Simplified Representation*, 1989.
ISO 8826–2:1994, *Technical Drawings – Roller Bearings – Part 2: Detailed Simplified Representation*, 1994.
ISO 9222–1:1989, *Technical Drawings – Seals for Dynamic Application – Part 1: General Simplified Representation*, 1989.
ISO 9222–2:1989, *Technical Drawings – Seals for Dynamic Application – Part 2: Detailed Simplified Representation*, 1989.
ISO 15786:2001, *Technical Drawings – Simplified Representation and Dimensioning of Holes*, 2001.

4

Dimensions, Symbols and Tolerances

4.0 Introduction

Dimensioning is necessary to define the shape and form of an engineering component. The basic principle of dimensioning has been covered already in Chapter 3 with respect to line types and thicknesses. This chapter continues the subject of dimensioning but considers some of the more fundamental principles of dimensioning and the implications with regard to inspection.

One of the problems of manufacture is that nothing can ever be made exactly to a size, even to atomic proportions. The surface of any component, even the smoothest, will vary. Thus, an inherent part of dimensioning must be a definition of the allowable variation. The permissible variation of a dimension is termed the *tolerance*. So, not only must dimensions be defined on a drawing but also tolerances.

4.1 Dimension definitions

When a part is to be dimensioned, the value and importance of any dimension will depend upon a variety of factors. These factors will be concerned with such things as the precision needed, the accepted variability, the function and the relationship to other features. In order to dimension correctly, a series of questions will need to be asked about a particular dimension and these will be dependent upon a clear understanding of the terminology associated with dimensions and tolerances. The following dimension definitions are important.

4.1.1 Functional and non-functional dimensions

Although every aspect of a component has to be dimensioned, some dimensions are naturally more important than others. Some dimensions will be critical to the correct functioning of the component and these are termed *functional dimensions*. Other dimensions will not be critical to correct functioning and these are termed *non-functional dimensions*. Functional dimensions are obviously the more important of the two and therefore will be more important when making decisions about the dimension value.

Figure 4.1 shows an assembly of a shaft, pulley and body. A shaft is screwed into some form of body and a pulley is free to rotate on the shaft in order to provide drive power via a belt (not shown). The details of the three parts of this assembly are shown below the assembly drawing. The important function dimensions are labelled 'F', and the non-functional dimensions 'NF'. The main function of the assembly is to allow the pulley to rotate on the shaft, driven by the belt. Thus, the bearing diameter and length of the bolt pulley are important and therefore they are functional dimensions because

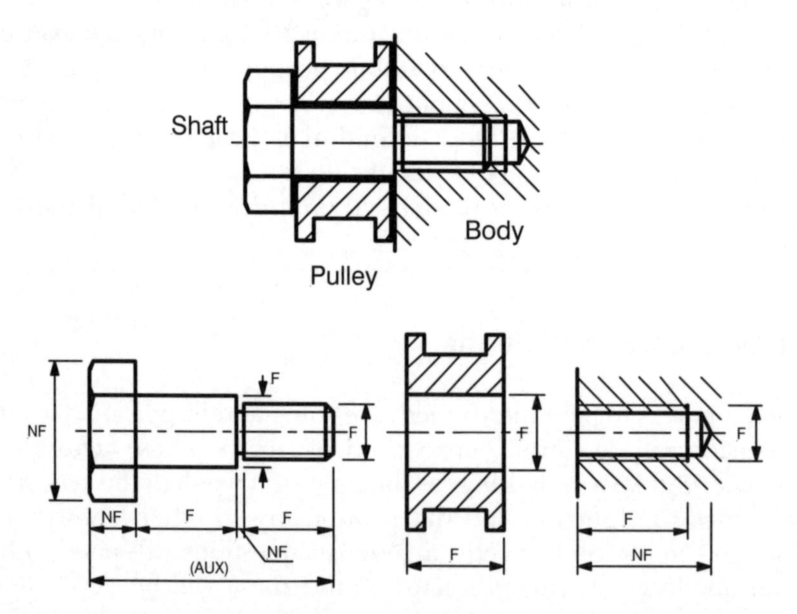

Figure 4.1 *Functional and non-functional dimensions of a pulley system*

they define the clearances that allow the pulley to rotate on the shaft. The belt will be under tension and the resulting lateral drive force will be transmitted to the shaft. The stresses set up by this force must be resisted by the screw thread in the body. Therefore, the length of engagement of the thread in the body is a functional dimension.

4.1.2 Auxiliary dimensions

The standard ISO 129:1985 states 'each feature shall be dimensioned only once on a drawing'. However, there are instances when there is a need for something to be dimensioned twice for information purposes. An example is shown in Figure 4.1. The bolt is made up of three sections, the head, the bearing shaft and the screw thread. Each has a length dimension associated with it. Together these make the shaft total length. It could be convenient, during manufacture, for the machinist to know the overall length of the bolt. If so, it could be provided as an *auxiliary dimension*. Auxiliary dimensions are in parentheses and no tolerances are added.

4.1.3 Features

Features are those aspects of a component which have individual characteristics and which need dimensioning. Examples of features are a flat surface, a cylindrical surface, a shoulder, a screw thread, a slot and a profile.

With respect to the pulley shown in Figure 4.1, there are eight features. These are the front and back annular flat surfaces, the front and back internal annular surfaces (in which the pulley runs), the two outside diameters of the flanges, the smaller outside diameter and the internal diameter. In a detailed drawing of the pulley all these eight features would need dimensioning. With respect to the tapped hole in the body in Figure 4.1, there are three features. These are the drilled hole, the threaded hole and the front flat surface.

4.1.4 Datums

A *datum* is a point, line or surface of a component to which dimensions are referred and from which measurements are taken during inspection. The datum point or points on a component are in

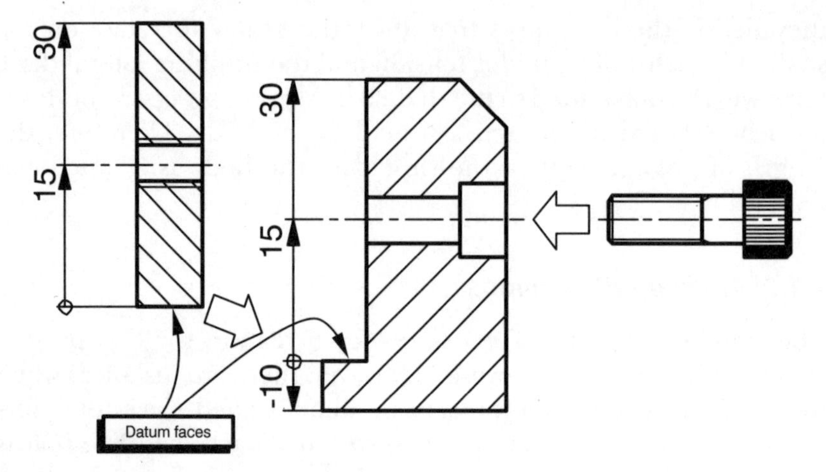

Figure 4.2 *Datum faces and dimensions*

reality datum features since it is with respect to features that other features relate. Thus, the various features on a component are dimensioned with respect to datum features. The definition of a datum feature will be dependent upon the functional performance of a particular component. In the example shown in Figure 4.2, the datum faces are the flat faces where the underside of the hardened insert contacts the step in the movable jaw. The 15mm dimension to the centre line of the bolt and the 30mm dimension to the top of the movable jaw are dimensioned from the datum surface. The underside of the movable jaw is 10mm from the datum surface but it is in the opposite direction and hence is labelled '–10'.

4.2 Types of dimensioning

There are essentially two methods of dimensioning features. Firstly, there is the addition of numerical values to dimension lines and secondly, there is the use of symbols. The first type has four elements to it and is the more usual method of dimensioning. This type has been described in Chapter 3 along with the relationship to the ISO rules. The second type uses symbols. In this case, a leader line with an arrowhead touches the feature referred to. Some form of symbol is placed at the other end of the leader line. These two types of dimensioning are described below.

4.2.1 Linear and angular dimensioning

The latest ISO standard concerned with dimensioning is ISO 129:1985. However, it is known that a new one is soon to be published which is ISO 129 Part 1 (the author sits on BSI/ISO committees and has seen the draft new standard). It has been through all the committee approval stages and has been passed for publication. However, it is being held back, awaiting the approval of a Part 2 so that they can be published together. The BSI estimates the publication date will be 2003, hence it will be ISO 129–1:2003. There have been versions prior to 1985 and each has defined slightly different dimensioning conventions. Needless to say, the 2003 convention gives a slightly different convention to the 1985 one! Throughout this section, the 2003 convention will be presented so that readers are prepared for the latest version.

Figure 4.3 shows a hypothetical spool valve that is defined by 14 dimensions in which 12 are linear and two angular. The valve is shown using the ISO principles of line thickness described in Chapter 3. Note that the valve outline uses the ISO type 'A' thickness whereas the other lines (including the dimension lines) are the ISO type 'B' thin lines. The outline thus has more prominence than the other lines and hence the valve tends to jump out of

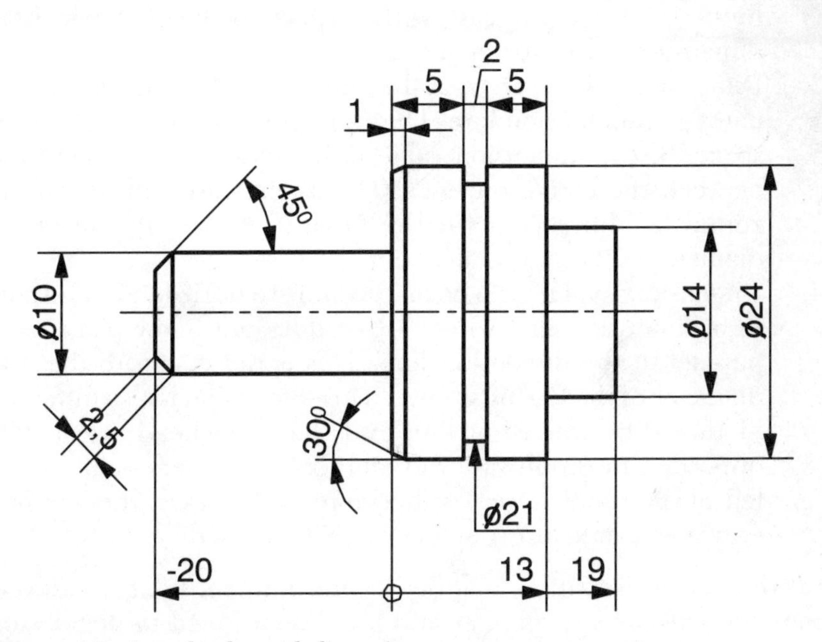

Figure 4.3 *A spool valve with dimensions*

the drawing page and into the eye of the reader. The valve dimensions follow the dimensioning convention laid down in the future ISO 129–1:2003 standard. Tolerances have been left off the figure for convenience. In this case there are two datum features. The first is the left-hand annular face of the largest cylindrical diameter, i.e. the face with the 30° chamfer. Horizontal dimensions associated with this datum face use a terminator in the form of a small circle. The other datum feature is the centre rotational axis of the spool valve represented by the chain dotted line. All the extension lines touch the outline of the spool valve. The dimension values are normally placed parallel to their dimension line, near the middle, above and clear of it. Dimension values should be placed in such a way that they are not crossed or separated by any other line.

There are several exceptions to this as seen in the drawing in Figure 4.3:

1. Dimension values of the running dimensions are shown close to the arrowheads and not in the centre of the dimension line. This applies to the '19', '13' and '–20' horizontal dimension, i.e. any running dimension value.
2. Dimension values can be placed above the extension of the dimension line beyond one of the terminators if space is limited. This is the case with respect to the '1' horizontal dimension of the 30° chamfer.
3. Dimension values can be at the end of a leader line that terminates at a dimension line. This applies when there is too little space for the dimension value to be added in the usual way between the extension lines. This is the case with the horizontal '2' dimension for the O-ring groove on the outer diameter of the spool valve.
4. Dimensional values can be placed above a horizontal extension to a dimension line where space does not allow placement parallel to the dimension line. This is the case with the '21' diameter of the O-ring groove. This dimension is also different in that the dimension line and the arrowhead are in the opposite direction to the '14' and '24' diameters seen on the left of the spool valve. Furthermore, in this case, the line has only one terminator (the other one is assumed).

All the above descriptions apply to linear dimensions. However, some dimensions are angular and are dimensioned in degrees or radians. The dimensioning of angles is just as important as the dimensioning of linear dimensions if a component is to be dimensioned correctly for manufacture. Angular dimensions use the same four elements as described above for linear dimensions.

In the case of the spool valve in Figure 4.3, two of the 14 dimensions are angles. These are the angles of the two chamfers. The 45° dimension has the two arrowhead terminators on the inside of the dimension arc whereas the 30° dimension has the arrowheads on the outside of the angular dimension arc. The latter is used because space is limited.

All dimension values, graphical symbols and annotations should normally be positioned such that they can be read from the bottom and from the right-hand side of the drawing. These are the normal reading directions. However, in some instances, reading from the bottom-right is not always possible if the requirements stated above are to be met. With reference to Figure 4.3, it is not possible to meet these requirements with respect to the 45° chamfer and the 2.5 dimension value. In these cases, the reading direction is the left-hand side and the bottom of the drawing. Figure 4.4 shows the common positions of linear dimensions and angular dimension values as given in the latest ISO standard.

Figure 4.5 shows three different methods of dimensioning-related features. With parallel and running dimensioning, the position of the hole centres as well as the right-hand plate edge are related to the datum left-hand edge. The advantage of parallel and running dimensioning is that every feature is related back to the

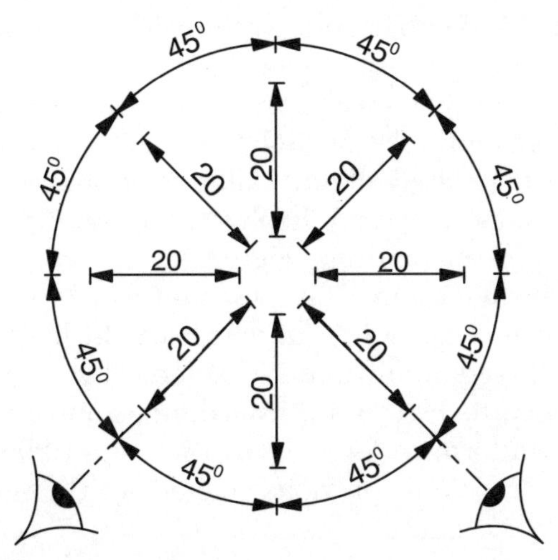

Figure 4.4 *Dimensioning different angular features*

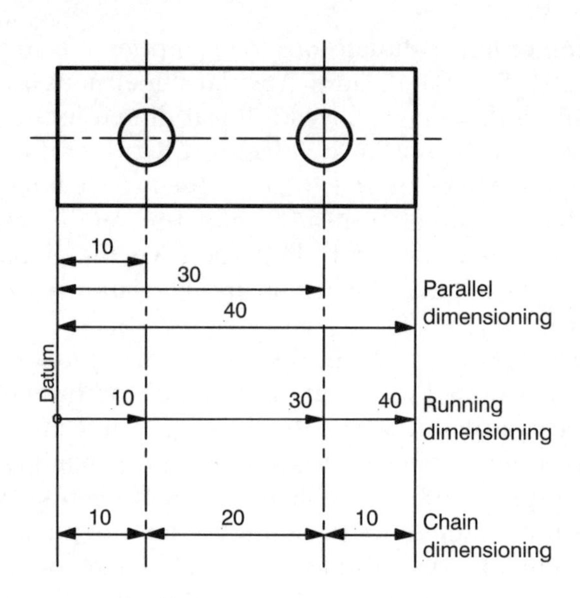

Figure 4.5 *Parallel, running and chain dimensioning*

same datum. Running and parallel dimensioning are identical methods. Chain dimensioning is an entirely different dimensioning methodology. In this case, only the left-hand hole is directly related to the left-hand datum surface. The right-hand hole is only related to the datum surface in a secondary manner and the right-hand edge is only related to the left-hand edge in a tertiary manner. The very name 'chain' illustrates the disadvantages in that the dimensions are chained together and not individually related back to a datum.

The dimensioning convention shown in Figure 4.3 is the convention given in the latest ISO 129–1:2003 standard. Previous ISO standards have used slightly different dimensioning conventions and of course it is very likely that some old drawings will conform to these conventions. Figure 4.6 shows a stepped shaft which has been dimensioned in a manner which is different from the convention in Figure 4.3 but which would have been recommended and allowed in previous ISO standards.

With reference to Figure 4.6, the dimensions on the drawing which do not conform to the current ISO 129–1:2003 convention but which do conform to previous versions of the standard are as follows:

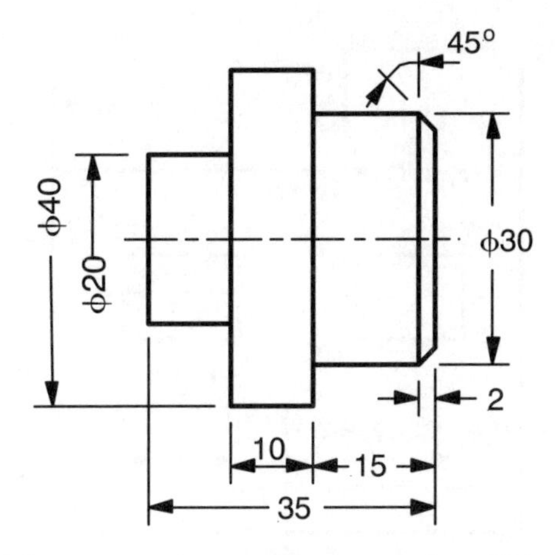

Figure 4.6 *Dimensioning practice according to previous ISO standards*

1. Projection lines do not touch the outside outline of the part. There is a small space between the part and the extension line. It makes the part stand-alone and away from the dimensioning and therefore easier to read.
2. Dimension values may be placed not parallel to the dimension line but perpendicular to it, e.g. the 'ϕ30'.
3. Dimension values may interrupt the dimension line, e.g. the 'ϕ30' and the '35'.
4. Diameters can be shown in half their full form, e.g. the 'ϕ20' and the 'ϕ40'.
5. Dimension values can be placed away from the centre of the dimension line and off to one side, e.g. the '45°' and the '2'.
6. Short dimensions using reversed arrows do not need to have a continuous dimension line, e.g. the '2' width of the chamfer.

4.2.2 Unacceptable dimensioning practice

It is a fundamental principle of dimensioning that the part and its dimensioning should be separated so that one does not impinge on the other and cause confusion. Such an example of incorrect dimensioning practice is shown in Figure 4.7. This is a plate with two circular holes and two rectangular holes in it. The various incorrect practices are shown by the word 'Wrong'. These are dimension lines being within the outline of the part, dimension lines crossing, broken extension lines, extension lines crossing dimensioning lines, shortened extension lines being used.

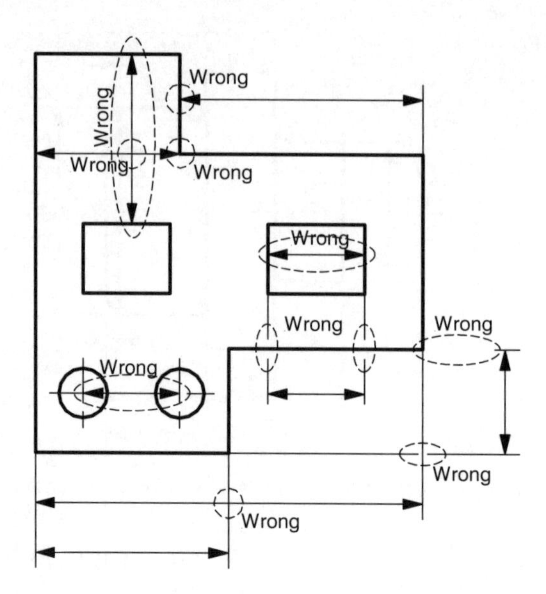

Figure 4.7 *Incorrect dimensioning practices*

4.3 Symbology

The ISO standards recommend that abbreviations and symbols are used wherever possible to avoid a link to any particular language. Examples of the use of symbology and English language abbreviations are as follows (BS 8888:2000):

'ϕ' or 'DIA' or 'D' or 'd' = diameter
'∩' = arc
'CL' = centre line
'CRS' = centres
'CSK' = countersunk
'CYL' = cylinder
'DRG' = drawing
'HEX' = hexagonal
'MMC' = maximum material condition
'PCD' = pitch circle diameter
'R' or 'RAD' = radius
'SP' = spherical diameter
'SQ' or □ (a small square) = a square feature
'SR' or 'ϕ' = spherical radius
'THD' = thread
'THK' = thick
'TOL' = tolerance
'VOL' = volume

Symbols can also be used to dimension holes. Figure 4.8 gives examples of the use of symbology with respect to specifying holes. Alongside these are standard dimensions. The first is an M10 threaded hole, second is a countersunk hole and the third is a stepped hole. If symbology is used, there is no need to add any dimensions using the conventional extension lines and arrowheads since symbols render dimensioning unnecessary. There is a whole

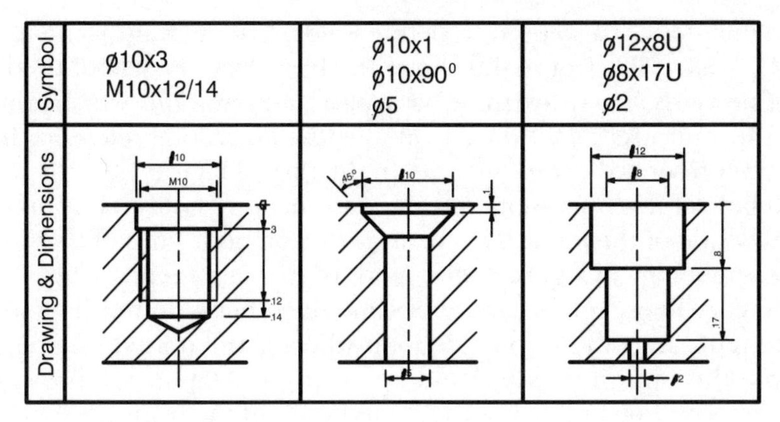

Figure 4.8 *Symbology for dimensioning holes*

ISO standard (ISO 15786:2003) devoted to the simplified representation and dimensioning of holes.

With regards to the movable jaw detailed drawing in Figure 3.2, there are four holes which are dimensioned using symbology. The meaning of each symbol is as follows.

The threaded hole has the symbology:

M8 × 10/12

This means that there is a single hole threaded to metric 8mm diameter which is 10mm deep. The full details of this metric thread are covered in the standard ISO 68–1:1998. The drilled hole, which was produced prior to tapping, is to be 12mm deep.

The single large counter-bored hole has the symbology:

ϕ15 × 7,5U
ϕ10

This means that there is one hole like this. The 15mm diameter portion of the hole is 7,5mm deep with a flat bottom (shown by the 'U') and the remainder is 10mm diameter.

The symbology for the two-off small counter-bored holes is as follows:

2 × ϕ8 × 5U
ϕ5

This means that there are two-off holes like this, hence the '2×' (two times). Both these holes have an upper portion which is 8mm in diameter and 5mm deep with a flat-bottomed hole (shown by the 'U'). The remainder of the hole is 5mm in diameter.

Symbology is also used to define welds. The relevant standard is ISO 2553:1992. Figure 4.9 shows the basic 'arrow' symbol used to define welds. There are three basic parts, an arrow line which points to the joint itself, a welding symbol and a horizontal reference line that represents the joint surface. In the case of Figure 4.9a, the weld symbol is for a fillet weld and there is only one weld that is on the arrow side of the joint. If there are welds on both sides of the joint, the symbology in Figure 4.9b is used. In this case there are two joint reference lines. One is continuous and the other is dotted. The solid line represents the arrow side joint whereas the dotted line represents the opposite side. Welding symbols are placed above the continuous line and below the dotted line. In many instances additional information is given, such as the weld dimensions, inspection rules and operating conditions like the welding rod specification. An example of additional information is shown in Figure 4.9c. The symbol in this case represents a low penetration single vee butt weld (the capital 'Y'). The 's5' refers to the fact that the weld depth is 5mm (see Figure 4.9d). The other numbers ('3 × 10(5)') mean that

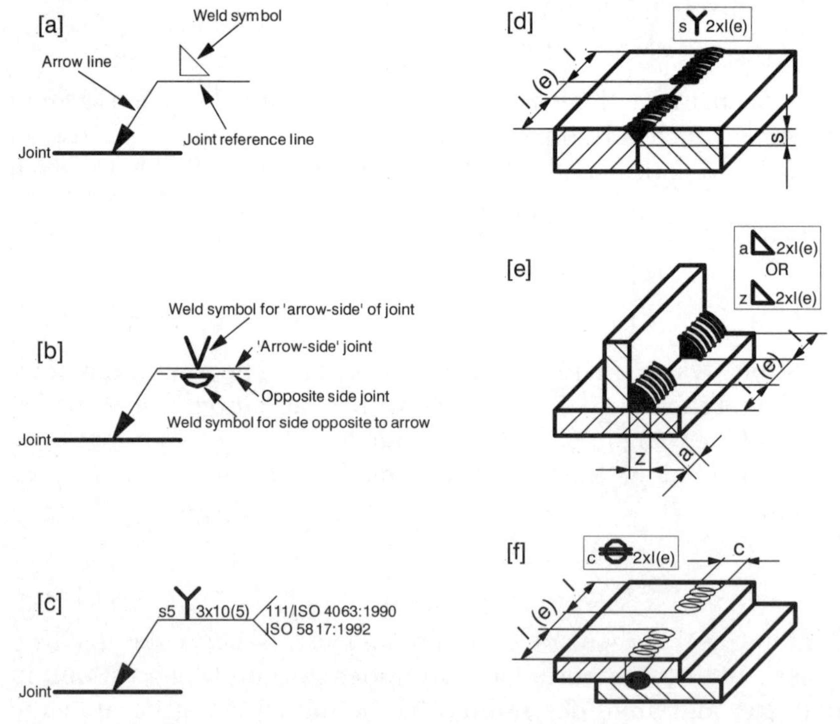

Figure 4.9 *Basic arrow symbol for representing welds*

there are 3 welds, each of 10mm length with a 5mm gap between them. Had it been a fillet weld, the starting letter would be either an 'a' or a 'z' (see Figure 4.9e). Had it been a seam weld, the starting letter would be a 'c' (see Figure 4.9f). In Figure 4.9c, the reference line has a fish tail end. Additional information is placed here. In this case the '111' is the code given in ISO 4063:1990 for metal-arc welding with a covered electrode. The next reference (ISO 5817:1992) refers to the acceptable weld quality level of imperfections. The ISO standard gives examples of other, highly specific information which can be referred to after the fish tail.

The table in Figure 4.10 shows numerous examples of welding symbology. The first column shows 10 types of weld and this is a selection from the 20 basic types. In each case the symbol used is representative of the actual weld. The symbol column shows two methods of representing the weld, one in which the arrow refers to the top face of the joint and one in which the arrow refers to the bottom face of the joint. In all cases there are two horizontal lines,

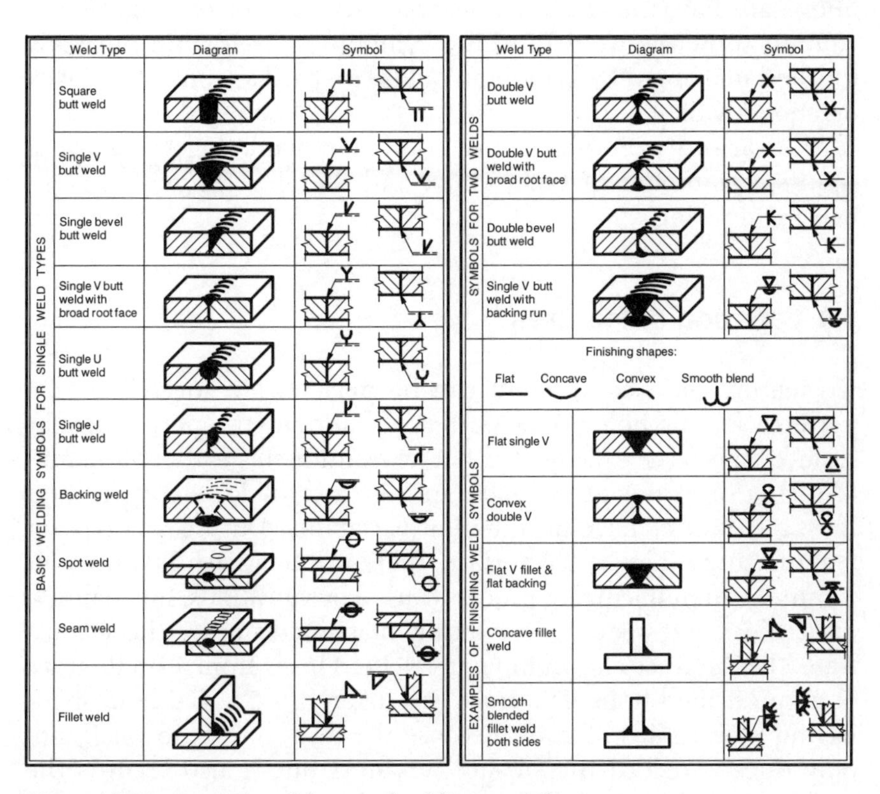

Figure 4.10 *Examples of the use of welding symbols*

one continuous and one dotted. In all instances the continuous line refers to the weld on the 'arrow face' of the joint and the dotted line to the opposite face. Considering the shallow penetration depth single vee butt weld of row 4, the welding symbol is 'Y'. When the arrow points to the top face, the 'Y' is on the continuous line and when the arrow points to the under side, the Y symbol is on the dotted line. Although it doesn't matter whether the continuous line is above or below the dotted, I think it is more logical for the symbols to actually reflect the layout of the weld itself. Therefore, I would suggest that when the arrow points to the upper face, the continuous line is above the dotted line whereas when the arrow points to the under side of the joint, it is more logical to have the dotted line above the continuous line.

With regard to the second column in Figure 4.10, the first four rows show how the previous set of symbols apply to welds on both sides of the joint. If the welds are the same, there is no need to have both a continuous line and a dotted line. In this case one continuous line is all that is necessary. The lower six rows of this right-hand column show the weld finishing symbology. Welds are normally finished machined by grinding. Welds can be flat, concave, convex or smoothly blended.

The various examples above show that symbology can save a significant amount of time, effort and therefore money in engineering drawing.

4.4 Variation of features

No feature on a component can be perfect. No surface can be perfectly flat, no hole can be perfectly round, no two perpendicular surfaces can be at exactly 90°. The reason for this is that all manufacturing processes are variable to a greater or lesser degree and thus, all features have an inherent variability. During any type of manufacture of, say, a flat surface, there will be variability inherent within the manufacturing process caused by vibrations, inequalities, instabilities and wear. For a typically machined surface this is illustrated by the trace shown in Figure 4.11. This is from a single trace of a gun-drilled hole. The trace was taken using a diamond stylus having a tip radius of 2um. Because the tip radius is so small, not only does it record the surface waviness but it also records the surface roughness caused by the individual machining scratches.

The trace is some 10mm long and it shows that the surface is not a 10mm long ideal straight line. The deviation over this 10mm length from the highest peak to the lowest valley is 4,2 microns yet this is a surface produced by precision machining.

It is not only flat surfaces that are variable. Figure 4.12 shows roundness traces from three positions along a ground hole. The traces do not indicate the diameter of the holes, merely their variability. The fact that they are three concentric circles of varying

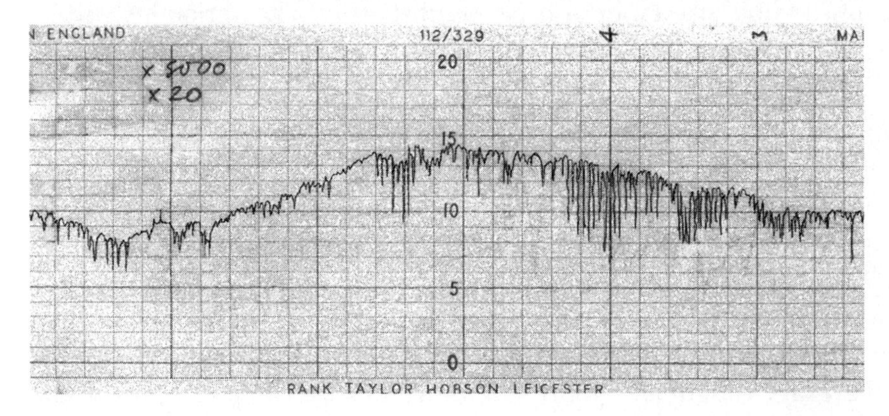

Figure 4.11 *Trace of a flat surface showing the deviations from the ideal straightness*

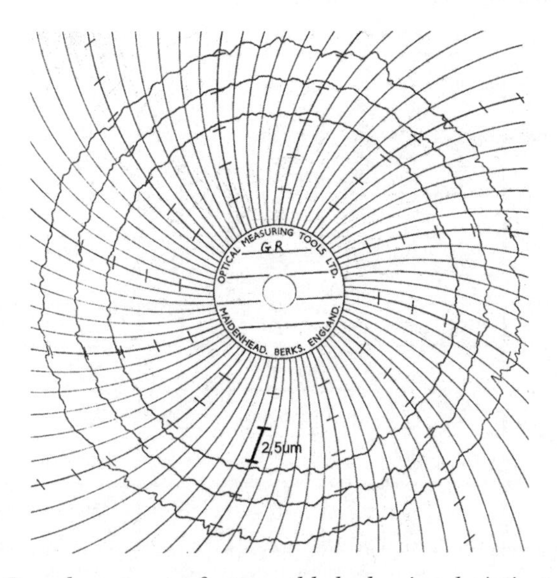

Figure 4.12 *Roundness traces of a ground hole showing deviations from an ideal circle*

diameter is due to the fact that the instrument settings are varied so that the radii can be separated. Each trace thus represents the circular trace around the ground bore and displays the out-of-roundness, not the absolute diameter. Clearly, each trace is far from an ideal circle, showing that even a precision ground hole has some variability.

The above two figures have demonstrated that a hole can never be perfectly straight or round. The same will apply to other aspects of the hole like taper and perpendicularity. The variability will be different each time a surface is produced on the same machine and also between different machines and processes. The variability will be higher with rough-machined surfaces and lower with precision-machined surfaces. The table in Figure 4.13 shows the variability of some hole manufacturing processes. The data refers to processes used for producing holes 25mm in diameter. In the figure, the word 'taper' means the maximum inclination over a 40mm length. The word 'ovality' means the difference between the maximum and minimum diameters at perpendicular positions. The word 'roundness' means the deviation from a true circle. The words 'average roughness' (represented by 'Ra', see Chapter 6) mean the average deviation of the surface micro-roughness after waviness has been removed. The table shows that on average, the variability for rough-machining processes is in the order of tens of microns

Data all for 25mm diameter holes	Drilling	Rough grinding	Rough boring	Reaming	Fine grinding	Honing	Broaching
Taper (um/40mm)	36	25	22	10	4	4	1
Ovality (um)	13	3	5	1	1	1	1
Roundness (um)	-	5	9	3	6	0.5	2
Average roughness (um) Ra	14	4,3	3,4	5,2	0,2	1,1	0,5
Cost relative to drilling	1	9	5	3	30	10	3

Figure 4.13 *Deviations, surface finishes and relative costs of 25mm diameter holes produced by a variety of manufacturing processes*

whereas the variability for precision-machined surfaces is in the order of microns. The table also shows the cost of producing the processes relative to drilling. In general, precision holes are more expensive to produce than rough-machined ones. One of the reasons for this is that higher quality machine tools are required to produce precision components. Typically, they would have more accurate bearings and have a more rigid and stable structure.

Figure 4.13 shows that holes can never be perfect cylinders. This then begs the question of what the real diameter of a hole is. The ovality shows that it varies in one direction in comparison to a perpendicular direction. The various drawings of components shown above (Figures 4.1, 4.2, 4.3 and 4.6) are therefore ideal representations of components since in reality all the component outlines drawn should be wavy lines since in reality there is always some variability. The result is that if one considers a hole, for example, it is impossible to state a single value for the diameter. However, it is possible to state maximum and minimum values that cover the range of the variability. Thus, when dimensioning any feature, two things must be provided: the basic nominal dimension and the permitted variability. This will be the nominal dimension plus a tolerance.

4.5 Tolerancing dimensions

There are essentially two methods of adding tolerances to dimensions: firstly universal tolerancing and secondly specific tolerancing. In the universal tolerance case, a note is added to the bottom of the drawing which says something like 'all tolerances to be ±0.1mm'. This means that all the features are to be produced to their nominal values and the variability allowed is plus or minus 0,1mm. However, such a blanket tolerance is unlikely to apply to each and every dimension on a drawing since some will be more important than others. Invariably, functional dimensions require a tighter (smaller) tolerance than non-functional dimensions.

A variation of universal tolerancing is where there are different classes of tolerance ranges applicable within a drawing. There are various ways of showing this on a drawing. One way is by the use of different numbers of zeros after the decimal marker. For example, a drawing may say:

'All tolerances to be as follows: *XX (e.g. 20) means ±0,5mm,*
XX,X (e.g. 20,0) means ±0,1mm
XX,XX (e.g. 20,00) means ±0,05mm'

In this case, any dimension on a drawing can be related to one of the three ranges given by the number of zeros used in the dimension value after the decimal marker.

The other method of dimensioning is specific dimensioning in which every dimension has its own tolerance. This makes every dimension and the associated tolerance unique and not related to any other particular tolerance, as is the case with general tolerancing. Figure 4.14 shows various ways of tolerancing dimensions. The first three are *bi-lateral tolerances* in that the tolerance is plus and minus about the nominal value whereas the last three are *uni-lateral tolerances* in that either the upper or the lower value of the tolerance is the same as the nominal dimension. The use of bi-lateral or uni-lateral tolerances will depend upon the tolerance situation and the functional performance. Note that, irrespective of whether bi-lateral or uni-lateral tolerancing is used, there are two general methods of writing the tolerances. The first is by putting the nominal value (e.g. 20) followed by the tolerance variability about that nominal dimension (e.g. +0,1 and –0,2). Alternatively, the maximum and minimum values of the dimension, including the tolerance can be given (e.g. 20,15 and 19,99). When dimensions are written down like this either as a tolerance about the nominal value or the upper and lower value method, the largest allowable dimension is placed at the top and the smallest allowable dimension at the bottom.

Normally, a mixture of general and specific tolerances is used on a drawing. The reason is that most dimensions are general and can be more than adequately covered by one or two tolerance ranges yet

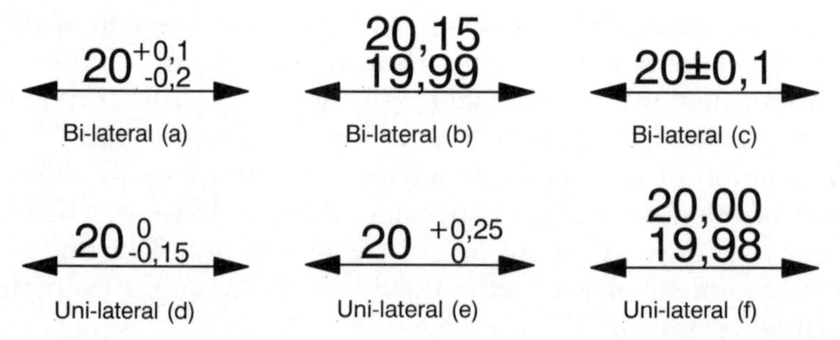

Figure 4.14 *The variety of ways that it is possible to add tolerances to a dimension*

there will be several functional dimensions that need specific and carefully described tolerance values. A good example of this would be the pulley bush in Figure 4.1. The bearing internal diameter tolerance would need to be tightly controlled to prevent vibration during high rotational speeds yet the outside diameter and the length could be defined by general tolerances.

Exactly the same principles apply to the dimensioning and hence tolerancing of angles. Indeed, the example shown in Figure 4.14 could just as easily have been drawn using angles as examples rather than linear measures.

Figure 4.5 has shown the difference between parallel, running and chain dimensioning. The important thing about parallel and running dimensions is that they are both related to a datum surface whereas this is not the case with chain dimensioning. When tolerances are added to parallel or running dimensions, the final variability result is significantly different from when tolerances are added to a chain dimension (see Figure 4.15). In the case of chain dimensioning, where each of the individual dimensions is cumulative, if tolerances are added to these dimensions, they too will be cumulative. This is not the case with running dimensions in that when a tolerance is applied to each running dimension the overall tolerances are the same for each dimension. In Figure 4.15, the three steps of the component are dimensioned using chain tolerancing (top) and running tolerancing (bottom). The shaded zones on the right-hand drawings show the tolerance ranges permitted by

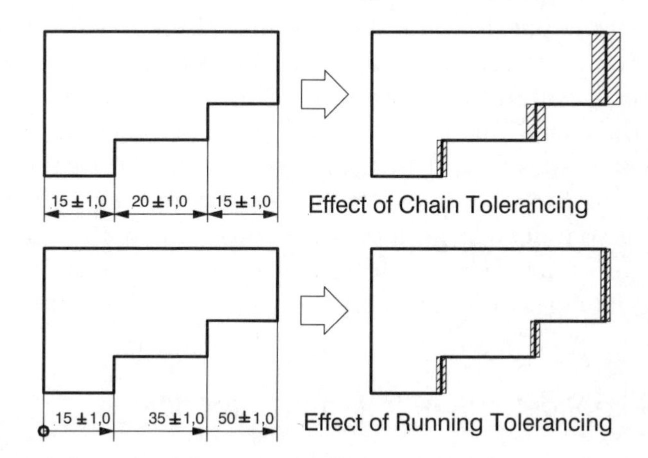

Figure 4.15 *The effect of different methods of tolerancing on the build-up of variability*

that particular method of dimensioning. In each case the tolerance on each dimension is ±1mm which is very large and only used for convenience of demonstration. Thus, with chain tolerancing, the final tolerance value at the end of the third step will be ±3mms whereas with running tolerances it will only be ±1mm.

4.6 The legal implications of tolerancing

The importance of correct tolerancing can be seen by the following example in which incorrect tolerancing resulted in a massive financial penalty for a company. A company produced a design drawing for a particular part which they sent out to a subcontractor for manufacture. The part was manufactured according to the drawings and returned to the contractor. Unfortunately, when the part was assembled into the main unit, it didn't fit. Some mating features did not align correctly and assembly was impossible. The contractor insisted the subcontractor had not made the part to the drawing and of course the subcontractor insisted they had! The case went to court and an expert witness was appointed. This expert witness was one of my predecessors in design teaching, hence I know about the case. The problem was that the designer in the contracting company used chain tolerancing when he should have used running tolerancing for a particular feature. He neglected to take into account the effect of tolerance build-up and the result was that the part did not fit in the assembly. Unfortunately, what he had in his mind he didn't put down on the drawing – back to communication 'noise' again (described in Chapter 1). The subcontractor made the part correctly within the chain tolerancing stated on the drawing so it wasn't their fault that the part didn't fit. The outcome of the case was that the court found in favour of the subcontractor and the contractor had to bear the costs. Such court and legal costs can be very high and indeed crippling. For example, in another case known by the author involving a design dispute, the court ruling and resulting damages were such that a subcontractor was bankrupted.

4.7 The implications of tolerances for design

The above explains the need for tolerances since nothing can be made perfectly. The following examples show how tolerances and clearances can be used together to make sure parts assemble. Figure 4.16 shows an example of the influence of hole clearances on position, dimensions and tolerances. The example consists of two plates bolted together. The top plate has two counter-bored clearance holes in it. The lower plate has two M5 threaded holes in it into which bolts are screwed. This example is concerned with the tolerance for the hole centre distance and the necessary clearances on the bolt in the upper plate. Let us assume that the hole spacing for the counter-bored holes in the top plate is invariant at 22,5mm. The tolerance associated with the threaded holes centre spacing in the lower plate is 22,5 ± 0,5mm. This tolerance of ±0,5mm is accommodated by the clearances on the bolt head and body of the counter-sunk holes in the top plate. These counter-sunk holes are over-sized to accommodate the hole centre spacing variability. The bolt shank diameter is 5mm and the head diameter is 8mm and the corresponding bolt hole diameters in the upper plate are 5,5mm and 8,5mm. This means that each bolt is 'free' to move ±0,25mm about the nominal value of 22,5mm to accommodate spacing variabilities.

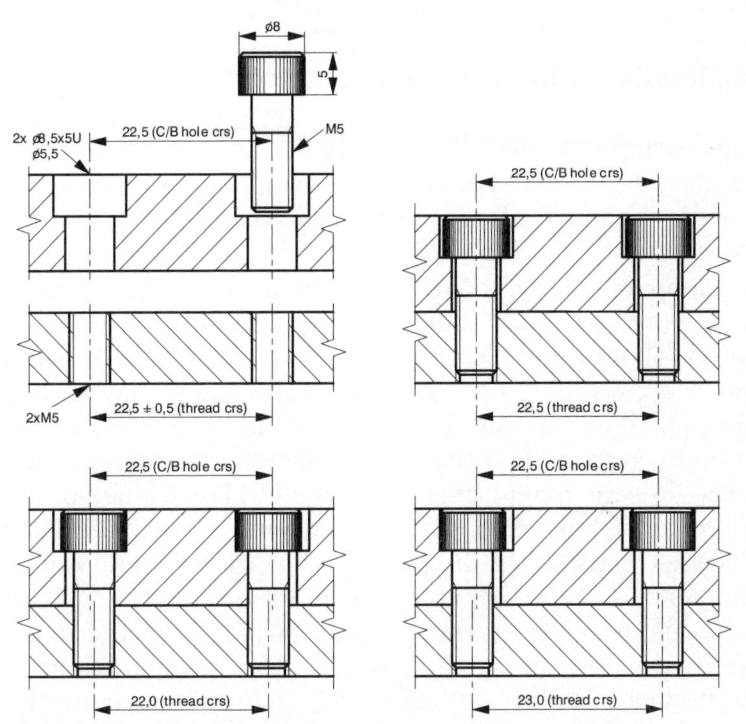

Figure 4.16 *The influence of hole clearances on hole centre position dimensions and tolerances*

The three small diagrams in Figure 4.16 show the three cases of nominal dimension, maximum dimension and minimum dimension. The top-right diagram shows the nominal situation where the threaded hole centre distance in the lower plate is the nominal value of 22,5mm. In this condition the bolts have an equi-spaced clearance on either side of the holes in the top plate. In the lower left-hand figure, the threaded holes centre distance is at the lowest value (i.e. 22,5 – 0,5 = 22,0mm). In this case the bolts and plates will still assemble because the clearances of the bolts in the upper plate allowed the bolts to be closer together. The lower right-hand figure case shows the situation when the threaded hole centre distance in the lower plate are in their maximum dimension condition (i.e. 22,5 + 0,5 = 23,00). In this case assembly is still possible because the clearances in the upper hole are such that the bolts can be positioned at their maximum spacing. It should be noted that the tolerance of 22,5 ± 0,5mm is a generous tolerance and has been given this value for convenience of drawing and understanding.

4.8 Manufacturing variability and tolerances

In the example shown in Figure 4.16, it was assumed that the holes and the bolts were all perfectly cylindrical and perfectly round. As has been explained above, this is not the case. The bolts and holes will all deviate from true circles due to manufacturing variabilities. An example of this is shown in Figure 4.17. This is a cross-section through the lower-right example in Figure 4.16. Here it can be seen that both bolts and holes deviate from circular. The deviation has been exaggerated for convenience of presentation and to make the point. The hole and bolt deviations are enclosed by maximum and minimum circles. The difference between the outer and inner circles gives the manufacturing variability. The contact position of the bolt in the hole will be given by the point at which the maximum enclosing diameter of the bolt touches the minimum enclosing diameter of the hole. The eccentricity created by this is shown by the equations of the diagram in Figure 4.17. Thus, the maximum permitted centre-line spacing of the holes (comparable to Figure 4.16 bottom-left diagram) will be the centre distance plus the two eccentricities. This is shown in the equation attached to Figure 4.17 and is the difference between the values of C(a) and C(b).

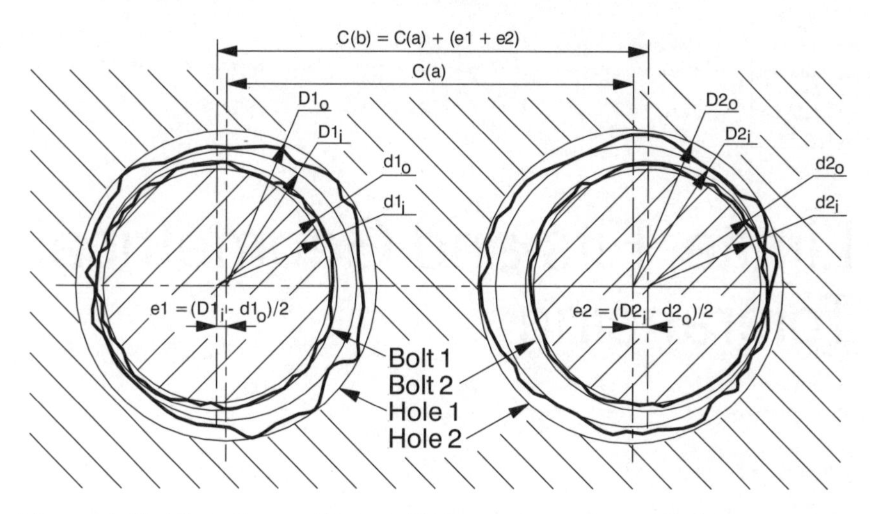

Figure 4.17 *The influence of bolt and hole out-of-roundness on hole centre position*

References and further reading

BS 8888:2000, *Technical Product Documentation – Specification for Defining, Specifying and Graphically Representing Products*, 2000.

ISO 68–1:1998, *General Purpose Screw Threads – Basic Profile: Part 1 – Metric Screw Threads*, 1998.

ISO 129:1985, *Technical Drawings – Dimensioning – General Principles, Definitions, Methods of Execution and Special Indications*, 1985.

ISO 129–1:2003, *Technical Drawings – Dimensioning – General Principles, Definitions, Methods of Execution and Special Indications*, 2003.

ISO 406:1987, *Technical Drawings – Tolerancing of Linear and Angular Dimensions*, 1987.

ISO 2553:1992, *Welded, Brazed and Soldered Joints – Symbolic Representation on Drawings*, 1992.

ISO 4063:1990, *Welding, Brazing, Soldering and Brazed Welding of Metals – Nomenclature of Processes and Reference Numbers for Symbolic Representation on Drawings*, 1990.

ISO 5459:1981, *Technical Drawings – Geometric Tolerancing – Datums and Datum Systems for Geometric Tolerancing*, 1981.

ISO 5817:1992, *Arc Welded Joints in Steel – Guidance on Quality Levels for Imperfections*, 1992.

ISO 15786:2003, *Technical Drawings – Simplified Representation and Dimensioning of Holes*, 2003.

5

Limits, Fits and Geometrical Tolerancing

5.0 Introduction

Previous chapters have underlined the importance of associating tolerances with dimensions because variability is always present. The question to be asked is how much variation is allowed with respect to functional performance and the selection of a manufacturing process. This is the subject of this chapter.

5.1 Relationship to functional performance

A journal bearing in a car engine is a convenient example of the necessity of carefully defining tolerances. If a journal bearing is designed to operate at high rotational speeds, the diamentral clearance is very important. If the clearance is too small, the bearing will seize whereas if the clearance is too large, the journal will vibrate within the bearing, creating noise, wear, vibration and heat. There is therefore an optimum clearance which is associated with smooth running. However, because variabilities are always present, an optimum range has to be specified rather than an absolute value. The left-hand drawing in Figure 5.1 shows a sketch of a journal bearing of nominal diameter 20mm, which has been designed to run at speed. The tolerances associated with the shaft and bearing are 19,959/19,980 and 20,000/20,033. These are the '*limits*' of size. They have been selected from special tables that relate certain performance situations to tolerance ranges (BS 4500A and B).

When the shaft and bearing are manufactured to these values the journal bearing will operate satisfactorily at speed without vibration or seizure. The tolerance ranges given in Figure 5.1 refer to a 'close-running fit'. The word '*fit*' is used specifically here because it describes the way that the journal fits in the bearing in terms of the dimensional relationships. For a 'close-running' fit, the tolerance ranges are given the designation: H8/f7. The standard tables show that the minimum diameter for the f7 shaft is 19,959mm and the maximum diameter is 19,980. With respect to the H8 hole, the minimum allowable diameter is 20,000mm and the maximum is 20,033. Thus, the average clearance is 47um, the minimum is 20um and the maximum is 74um. This means that if the clearance in the journal bearing is less than 20um, it will seize and if it is greater than 74um, wear and vibrations will result. Under these 'close-running fit' tolerances, the shaft and bearing will perform satisfactorily.

The right-hand sketch in Figure 5.1 shows a 'sliding fit'. This would apply to, say, a spool valve in which a shaft translates and/or rates at slow speed. The 'sliding fit' class corresponds to tolerance grades H7 and g6. The H7 tolerance applies to the hole and is 21um (i.e. 20,021–20,000). The shaft tolerance is g6 and is 13um (i.e. 19,993–19,980). These tolerance bands mean that the maximum clearance is 41um, the minimum clearance is 7um and

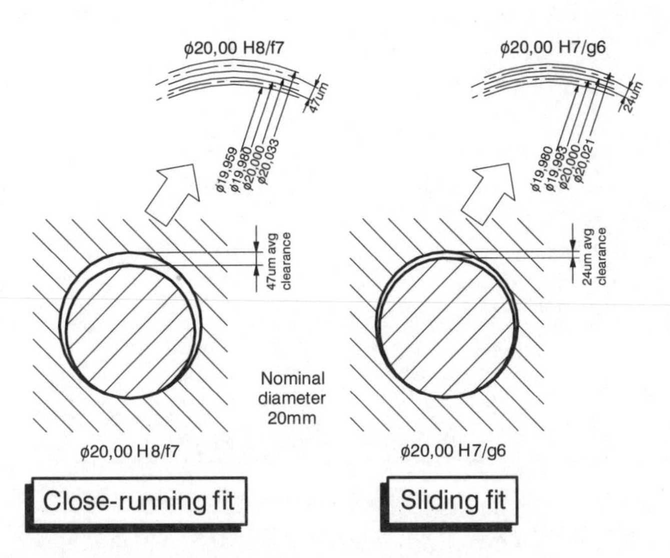

Figure 5.1 *Examples of two different types of bearings and their tolerances*

the average is 24um. These are about half the values of the 'close-running' fit of the left-hand sketch in Figure 5.1.

5.2 Relationship to manufacturing processes

In any machining process, the tolerance that can be achieved will depend upon two things. Firstly, the variability caused by the vagaries within a manufacturing process such as vibrations, discontinuities, inconsistencies, etc. These will produce a deviation about some mean value. Secondly, there is the variation that occurs when the tool wears. This will be progressive. Thus, in any accuracy graph or table, there will be two factors: an increasing trend with wear and variability scattered around this trend. This is shown in the graph in Figure 5.2. The nominal diameter was 10mm and the manufacturing process was gun-drilling. The graph shows that there is a general trend produced by wear and variability given by the 'error' bars essentially equi-spaced about the mean. In this case the variability about the mean value represents the out-of-roundness. This

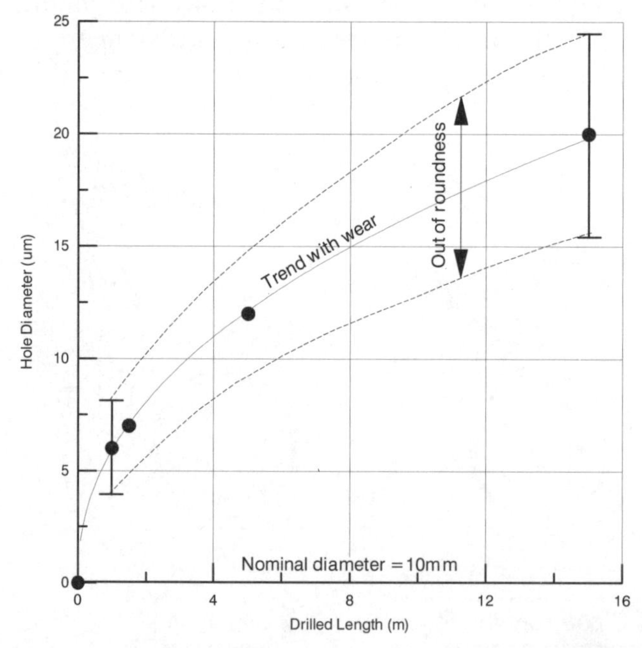

Figure 5.2 *Gun-drill wear against hole diameter showing wear trend and out-of-roundness*

is the deviation of the hole from a perfectly circular hole. The out-of-roundness refers to random as well as systematic errors.

An example of a systematic error is shown in the picture in Figure 5.3. This is a photograph of a 6mm-diameter hole in a 3mm thick aluminium sheet. The hole is clearly of a triangular form. The 'halo' round the edge of the hole is where it has been chamfered to remove the burr. The reason the hole is triangular is because of a lack of stability of the drill caused mainly by the fact that the tip breaks through the thin sheet before the outer edges are engaged in cut. The ensuing vibrations have caused the drill to both rotate and oscillate. It is significant that a 2-point measurement using, say, a digital calliper produces an almost constant diameter of 6,5mm whereas in fact the circumscribed circle diameter is some 15% larger than the inscribed circle diameter. This difference would be seen if a 3-leg internal micrometer were used to measure the hole.

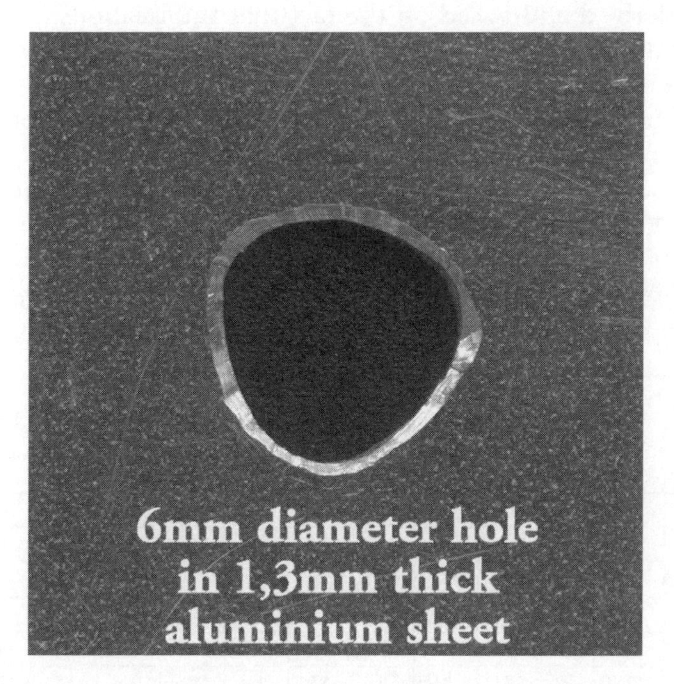

Figure 5.3 *A 6mm-diameter hole drilled in a thin aluminium sheet using a twist drill*

5.3 ISO tolerance ranges

Tolerance bands need to be defined which can be related to functional performance and manufacturing processes. The ISO has published tolerance ranges to help designers. Examples of these tolerance ranges are shown in Figure 5.4. This table is only a selection from the full table given in ISO 286–2:1988. The full range goes up to IT18 and 3m nominal size. The tolerance ranges are defined by 'IT' ranges as shown in the diagram from IT1 to IT11. The range given in the ISO standard is significantly more complicated than the extract in Figure 5.4. It should be noted that the range increases as the IT number gets larger and the range increases as the nominal size increases. The latter is fairly logical in that one would expect the tolerance range to be larger as the diameter increases because the precision that can be achieved must be relative. The ranges were not chosen out of the blue but empirically derived and based on the fact that the relationship between manufacturing errors and basic size can be approximated by a parabolic function.

The trace from a flat surface shown in Figure 4.11 has shown the maximum deviation over the 10mm length to be 4,2um. The nominal size was 22mm. If this surface was to be inspected with respect to the tolerance grades in Figure 5.4, the 22mm nominal size would fall within the row 18 to 30mm. Along this row, the 4,2um corresponds to IT4 since, if the tolerance on a drawing was given by IT1 to IT3, the surface would fail inspection whereas if the drawing

Nominal size		ISO Tolerance ranges in microns										
Over	Up to & incl	IT1	IT2	IT3	IT4	IT5	IT6	IT7	IT8	IT9	IT10	IT11
-	3	0,8	1,2	2	3	4	6	10	14	25	40	60
3	6	1	1,5	2,5	4	5	8	12	18	30	48	75
6	10	1	1,5	2,5	4	6	9	15	22	36	58	90
10	18	1,2	2	3	5	8	11	18	27	43	70	110
18	30	1,5	2,5	4	6	9	13	21	33	52	84	130
30	50	1,5	2,5	4	7	11	16	25	39	62	100	160
50	80	2	3	5	8	13	19	30	46	74	120	190
80	120	2,5	4	6	10	15	22	35	54	87	140	220
120	180	3,5	5	8	12	18	25	40	63	100	160	250
180	250	4,5	7	10	14	20	29	46	72	115	185	290

Figure 5.4 *Standard ISO tolerance ranges adapted from ISO 286–2:1988*

specified IT4 or above, it would pass the inspection. Similarly, with respect to the gun-drilled hole out-of-roundness deviation in Figure 5.2, the bars on the graph show that with a sharp drill, the out-of-roundness is 4,5um whereas when the drill is worn the out-of-roundness is 9,1um. These values beg the question as to what IT class this gun-drilling hole belongs to. The quick answer is that it depends on drill wear. With reference to Figure 5.4, the appropriate row is 6 to 10mm (i.e. the third row). The 4,5um out-of-roundness corresponds to class IT5 whereas the 9,1um out-of-roundness corresponds to class IT7. If the tolerance class IT4 is to be met by gun-drilling then a drill can only be used for a short proportion of its life. If, on the other hand, class IT7 is acceptable, this can be achieved throughout the life of the drill.

Figure 5.5 shows the IT tolerance ranges for various situations. These are the ranges for measuring tools, for common manufacturing processes, for limits and fits and for the production of materials. It is perhaps of no surprise that the range produced by common manufacturing processes is almost the same as the range of limits and fits from which designers can select functional performance tolerances.

Figure 5.6 is a table that is essentially an expansion of the manufacturing processes range in Figure 5.5. This table shows the range of tolerances achieved by the most common manufacturing processes. High-precision processes like lapping can achieve tolerance IT4 whereas, at the other end, roughing processes like shaping are only IT11. The range within any one process represents the variabilities caused by such things as wear, feed and speed, etc.

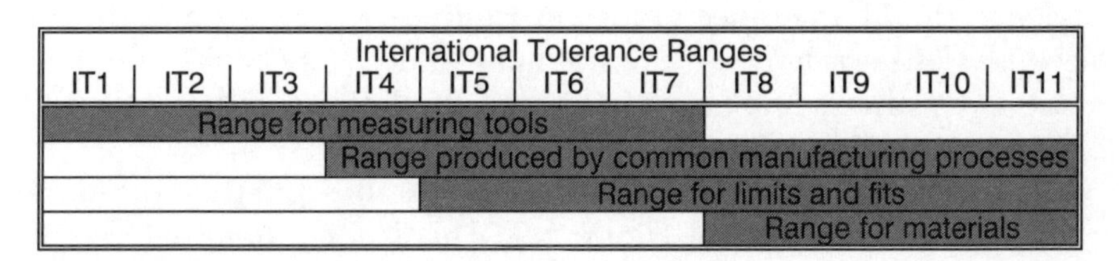

Figure 5.5 *ISO tolerance ranges for various situations*

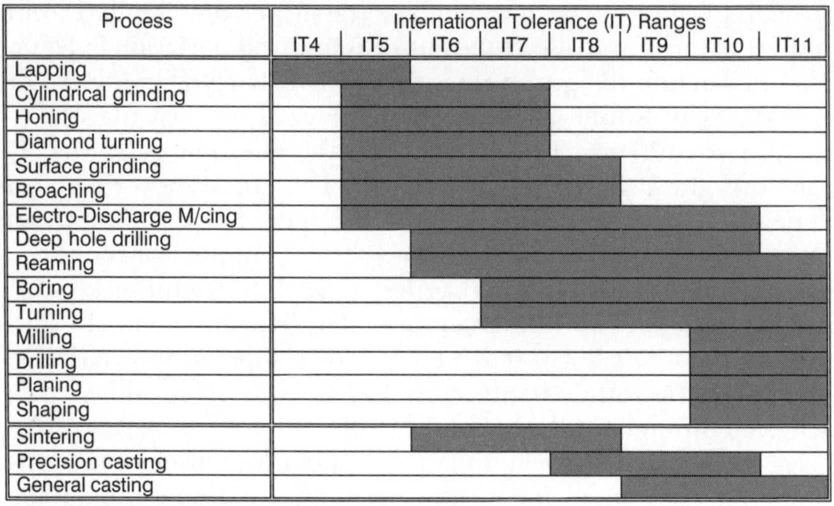

Process	International Tolerance (IT) Ranges							
	IT4	IT5	IT6	IT7	IT8	IT9	IT10	IT11
Lapping	■	■						
Cylindrical grinding		■	■	■				
Honing		■	■	■				
Diamond turning		■	■	■				
Surface grinding		■	■	■	■			
Broaching		■	■	■	■			
Electro-Discharge M/cing		■	■	■	■	■	■	
Deep hole drilling			■	■	■	■	■	
Reaming			■	■	■	■	■	■
Boring				■	■	■	■	■
Turning				■	■	■	■	■
Milling							■	■
Drilling							■	■
Planing							■	■
Shaping							■	■
Sintering			■	■	■			
Precision casting					■	■	■	
General casting						■	■	■

Figure 5.6 *ISO tolerance ranges for a variety of manufacturing processes*

5.4 Limits and fits

The tolerance ranges shown in Figures 5.4, 5.5 and 5.6 are simply ranges. To relate to function they must be put into context and related to some absolute datum. This is the situation demonstrated by the bearings in Figure 5.1. Considering the 'close-running fit' example, the tolerance ranges are IT8 for the hole and IT7 for the shaft. However, it is insufficient to just quote an IT tolerance class on its own. The tolerance class must be related to a datum, in this case the nominal 20mm diameter. The shorthand way of referring to these limits is the designations 'H8' and 'f7'. The '8' and the '7' refer to the IT tolerance grades in Figure 5.4. The 'H' and the 'f' give the offset relative to the nominal value. Note that the upper case letter always applies to holes and the lower case letter always applies to shafts.

The relationship between the tolerance grades and their offsets is shown in the diagram in Figure 5.7. This is for a nominal size of 25mm diameter and tolerance range IT7. Shaft tolerance ranges are represented by the lower-case letters a to z and holes by the upper-case letters A to Z. Since these are all for the ISO tolerance range IT7, the values should be a7 to z7 and A7 to Z7 respectively. Note that the two sets of bars in Figure 5.7 (for holes and shafts) are the inverse of each other.

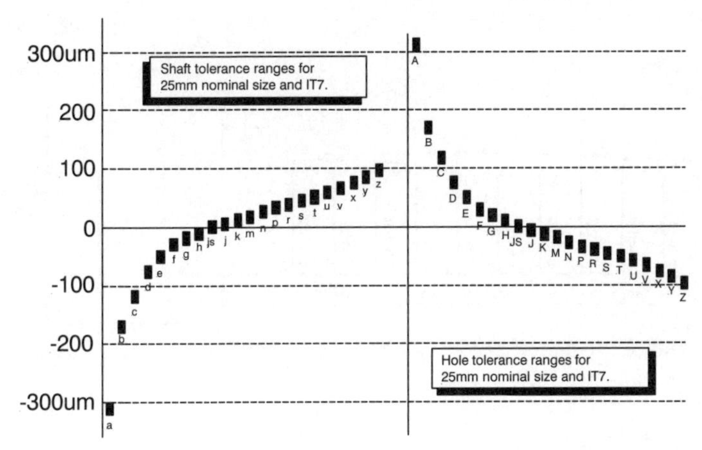

Figure 5.7 *ISO shaft and hole tolerance classes for 25mm nominal size and range IT7*

The alphanumeric tolerance range classes typified in Figure 5.7 can be used to inspect components produced by manufacturing processes. As an example, let us assume we want to inspect a shaft which is to be a 'close-running fit' in a journal as per the left-hand diagram in Figure 5.1. The shaft would be represented by the designation ϕ20,00 f7. The upper size limit for class f7 is 19,980mm diameter and the lower size limit for class f7 is 19,959mm diameter. If the shaft were produced on a lathe, there will be a size variability which depends upon the operating conditions and the tool wear. We need to reject any shafts that have a diameter in excess of the upper size limit as well as those which have a diameter that is lower than the lower size limit. This would ensure that the only turned shafts that pass the inspection process are those which meet the requirements if the class is f7. Such an inspection situation is demonstrated by the schematic diagram in Figure 5.8. The basic inspection device is a 'go/no-go' gauge which has one recess corresponding to the upper size limit and another recess which corresponds to the lower size limit for class f7. In this case we are assuming that 10 shafts are manufactured and each is inspected using the go/no-go gauge. To pass inspection, each must be able to enter the left-hand 'go' gauge but not the right-hand 'no-go' gauge. Assuming that the sizes for the 10 shafts are as shown, shafts 1, 2, 3, 4, 7, 8, 9 and 10 pass the f7 inspection test whereas shafts 5 and 6 are rejected because they are undersized and oversized respectively.

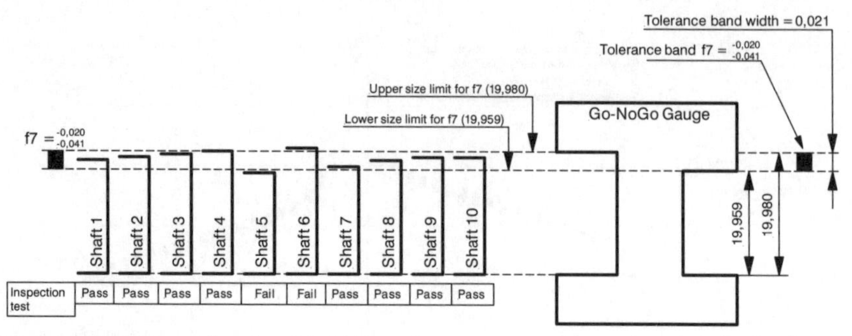

Figure 5.8 *Example of a 20,00f7 go/no-go gauge inspecting 10 shafts from a production line*

5.4.1 Fit systems

Figure 5.9 shows the three basic fit 'systems'. The left-hand sketch shows a shaft which will always fit in the hole because the shaft maximum size is always smaller than the hole minimum size. This is called a *clearance fit*. These have been discussed above with respect to running and sliding fits as per Figure 5.1. In some functional performance situations, an *interference fit* is required. In this case, the shaft is always larger than the hole. This would be the case for the piston rings prior to their assembly within an engine bore or for a hub on a shaft. In some functional performance situations, a *transition fit* may be required. Should the shaft and hole final diameters be an interference-clearance fit, the clearances will be very small and the location would be very accurate. If it were an interference-transition fit, on assembly the shaft would 'shave' the hole and thus the location would be very accurate.

5.4.2 The 'shaft basis' and the 'hole basis' system of fits

In all the examples given above, the discussion has been concerning 'shafts' and 'holes'. It should be remembered that this does not necessarily apply to shafts and holes. These are just generic terms that mean anything that fits inside anything else. However, whatever the case, it is often the case that either the shaft or the hole is the easier to produce. For example, if they are cylindrical, the shaft will be the more easily produced in that one turning tool can produce an infinite number of shaft diameters. This is not the case with the cylindrical hole in that each hole size will be dependent on a single drill or reamer.

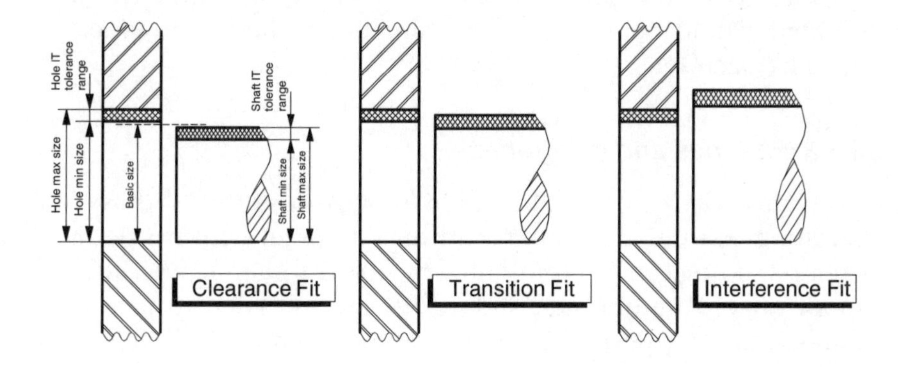

Figure 5.9 *Typical clearance, transition and interference fits for a shaft in a hole*

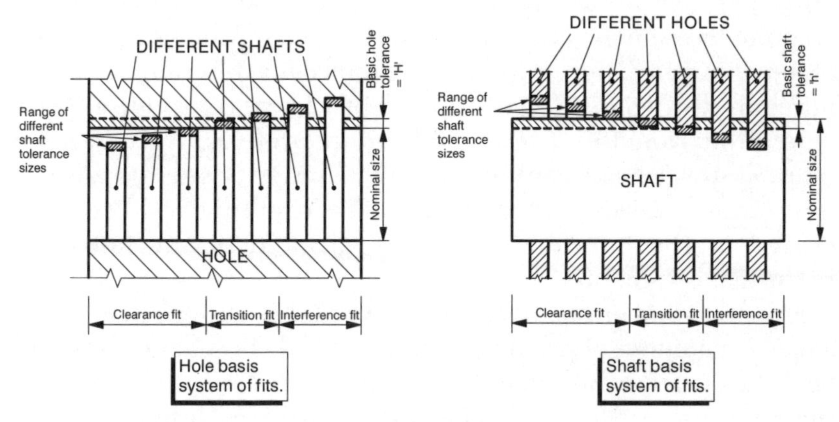

Figure 5.10 *Hole basis and shaft basis examples of fits*

The right-hand diagram in Figure 5.10 shows the situation in which the shaft is the more difficult of the two to produce and this is referred to as the 'shaft basis' system of fits. In this case the system of fits is used in which the required clearances or interferences are obtained by associating holes of various tolerance classes with shafts of a single tolerance class. Alternatively if the shaft is the easier part to produce then the hole basis system of fits is used. This is a system of fits in which the required clearances and interferences are obtained by associating shafts of various tolerance classes with holes of a single tolerance class. In the case of the shaft basis system the shaft is kept constant and the interference or clearance functional situation is achieved by manipulating the hole. If the hole-based system is used, the opposite is the case. The appropriate use of each system for functional performance situation is thus made easier for the manufacturer.

5.4.3 Fit types and categories

Clearance fits can be subdivided into running or sliding fits. Running applies to a shaft rotating at speed within a journal whereas sliding can be represented by slow translation, typically of a spool valve. Running and sliding fits are intended to provide a similar running performance with suitable lubrication allowance throughout a range of sizes. Transition fits are used for locational purposes. Because of the difference in sizes they will either be low clearance fits or low interference fits. They are intended to provide only the location of mating parts. They may provide rigid or accurate location as with interference fits or provide some measure of freedom in location as in small clearance fits. Interference fits are normally divided into force or shrink fits. These constitute a special type of interference. The idea of the interference is to create an internal stress that is constant through a range of sizes because the interference varies with diameter. The resulting residual stress caused by the interference will be dictated by the functional performance situation.

From the data given above it should be fairly obvious that there is a massive number of permutations of fits and classes and sizes. This begs the question, how does a designer select a particular one from the multitude available? The answer is that designers use a preferred set of fits. Many examples of preferred fits are available. Examples of commonly used ones are given in the standards BS 4500A and BS 4500B re British practice. The charts of preferred fits given in Figures 5.11 and 5.12 are a subset of the BS 4500 selection. Although these eight classes are just a selection, they represent archetypal cases. Regarding clearance fits, the loose running fit class is for wide commercial tolerances or allowances. The free running fit is not for use where accuracy is essential but is appropriate for large temperature variations, high running speeds or heavy journal pressures. The close running fit is for running on accurate machines or for accurate location at moderate speeds and journal pressures. The sliding fit is not intended to run freely but to move and turn freely and locate accurately. The low locational transition fit is for accurate location and is a compromise between clearance and interference. The high locational transition fit is for more accurate location where greater interference is permissible. The locational interference fit is for parts requiring rigidity and alignment with the prime accuracy of location but without any special residual pressure requirement. The medium drive fit is for ordinary steel parts or shrink fits on light sections. It is the tightest fit useable with cast iron. These eight classes provide a useful starting point for most functional performance situations.

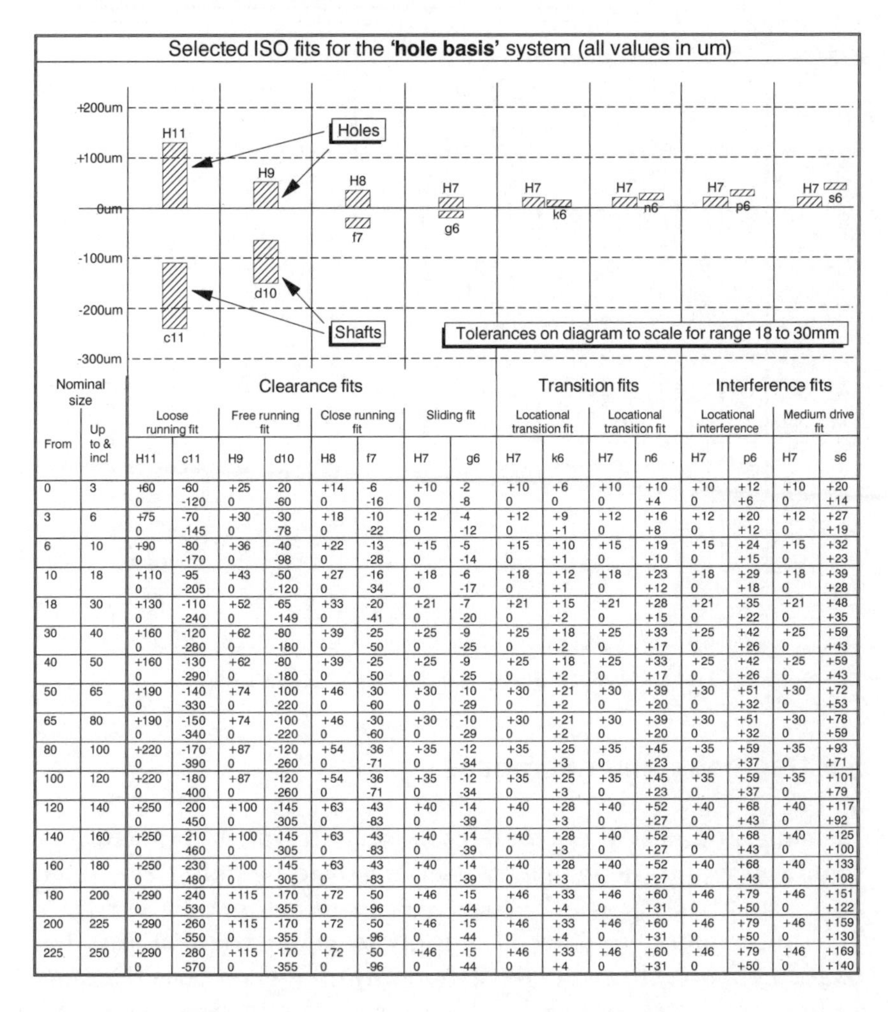

Nominal size		Clearance fits								Transition fits				Interference fits			
		Loose running fit		Free running fit		Close running fit		Sliding fit		Locational transition fit		Locational transition fit		Locational interference		Medium drive fit	
From	Up to & incl	H11	c11	H9	d10	H8	f7	H7	g6	H7	k6	H7	n6	H7	p6	H7	s6
0	3	+60 0	-60 -120	+25 0	-20 -60	+14 0	-6 -16	+10 0	-2 -8	+10 0	+6 0	+10 0	+10 +4	+10 0	+12 +6	+10 0	+20 +14
3	6	+75 0	-70 -145	+30 0	-30 -78	+18 0	-10 -22	+12 0	-4 -12	+12 0	+9 +1	+12 0	+16 +8	+12 0	+20 +12	+12 0	+27 +19
6	10	+90 0	-80 -170	+36 0	-40 -98	+22 0	-13 -28	+15 0	-5 -14	+15 0	+10 +1	+15 0	+19 +10	+15 0	+24 +15	+15 0	+32 +23
10	18	+110 0	-95 -205	+43 0	-50 -120	+27 0	-16 -34	+18 0	-6 -17	+18 0	+12 +1	+18 0	+23 +12	+18 0	+29 +18	+18 0	+39 +28
18	30	+130 0	-110 -240	+52 0	-65 -149	+33 0	-20 -41	+21 0	-7 -20	+21 0	+15 +2	+21 0	+28 +15	+21 0	+35 +22	+21 0	+48 +35
30	40	+160 0	-120 -280	+62 0	-80 -180	+39 0	-25 -50	+25 0	-9 -25	+25 0	+18 +2	+25 0	+33 +17	+25 0	+42 +26	+25 0	+59 +43
40	50	+160 0	-130 -290	+62 0	-80 -180	+39 0	-25 -50	+25 0	-9 -25	+25 0	+18 +2	+25 0	+33 +17	+25 0	+42 +26	+25 0	+59 +43
50	65	+190 0	-140 -330	+74 0	-100 -220	+46 0	-30 -60	+30 0	-10 -29	+30 0	+21 +2	+30 0	+39 +20	+30 0	+51 +32	+30 0	+72 +53
65	80	+190 0	-150 -340	+74 0	-100 -220	+46 0	-30 -60	+30 0	-10 -29	+30 0	+21 +2	+30 0	+39 +20	+30 0	+51 +32	+30 0	+78 +59
80	100	+220 0	-170 -390	+87 0	-120 -260	+54 0	-36 -71	+35 0	-12 -34	+35 0	+25 +3	+35 0	+45 +23	+35 0	+59 +37	+35 0	+93 +71
100	120	+220 0	-180 -400	+87 0	-120 -260	+54 0	-36 -71	+35 0	-12 -34	+35 0	+25 +3	+35 0	+45 +23	+35 0	+59 +37	+35 0	+101 +79
120	140	+250 0	-200 -450	+100 0	-145 -305	+63 0	-43 -83	+40 0	-14 -39	+40 0	+28 +3	+40 0	+52 +27	+40 0	+68 +43	+40 0	+117 +92
140	160	+250 0	-210 -460	+100 0	-145 -305	+63 0	-43 -83	+40 0	-14 -39	+40 0	+28 +3	+40 0	+52 +27	+40 0	+68 +43	+40 0	+125 +100
160	180	+250 0	-230 -480	+100 0	-145 -305	+63 0	-43 -83	+40 0	-14 -39	+40 0	+28 +3	+40 0	+52 +27	+40 0	+68 +43	+40 0	+133 +108
180	200	+290 0	-240 -530	+115 0	-170 -355	+72 0	-50 -96	+46 0	-15 -44	+46 0	+33 +4	+46 0	+60 +31	+46 0	+79 +50	+46 0	+151 +122
200	225	+290 0	-260 -550	+115 0	-170 -355	+72 0	-50 -96	+46 0	-15 -44	+46 0	+33 +4	+46 0	+60 +31	+46 0	+79 +50	+46 0	+159 +130
225	250	+290 0	-280 -570	+115 0	-170 -355	+72 0	-50 -96	+46 0	-15 -44	+46 0	+33 +4	+46 0	+60 +31	+46 0	+79 +50	+46 0	+169 +140

Figure 5.11 *Eight archetypal fits for the 'hole basis' system of fits*

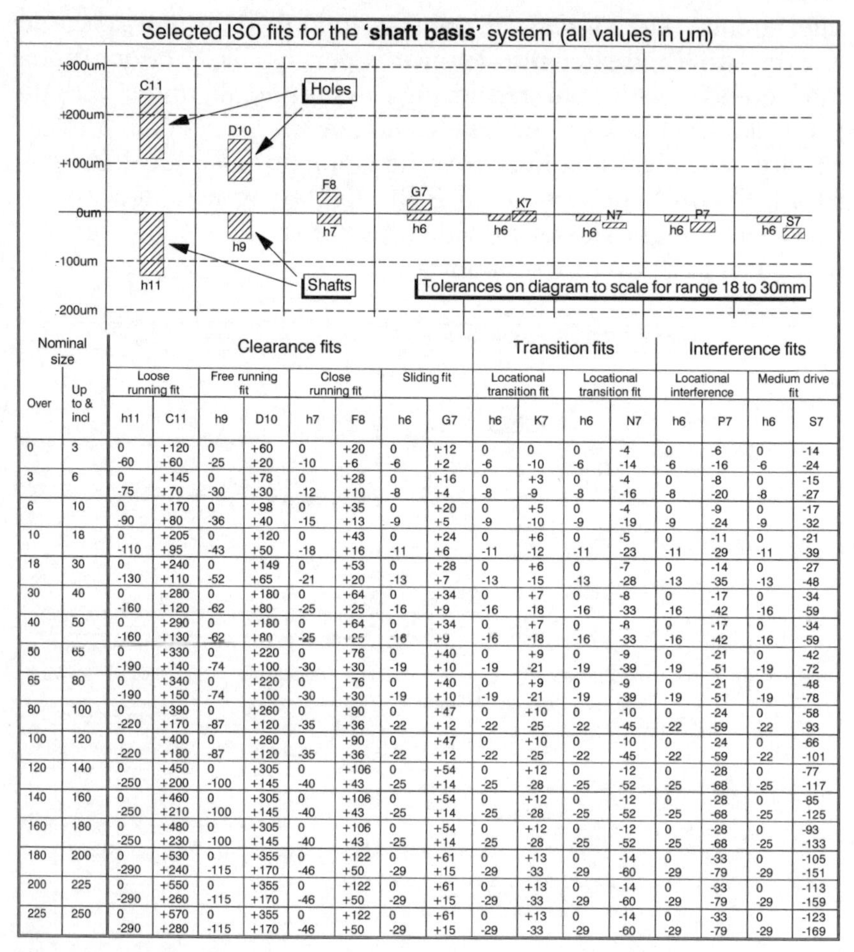

Nominal size		Clearance fits								Transition fits				Interference fits			
		Loose running fit		Free running fit		Close running fit		Sliding fit		Locational transition fit		Locational transition fit		Locational interference		Medium drive fit	
Over	Up to & incl	h11	C11	h9	D10	h7	F8	h6	G7	h6	K7	h6	N7	h6	P7	h6	S7
0	3	0 -60	+120 +60	0 -25	+60 +20	0 -10	+20 +6	0 -6	+12 +2	0 -6	0 -10	0 -6	-4 -14	0 -6	-6 -16	0 -6	-14 -24
3	6	0 -75	+145 +70	0 -30	+78 +30	0 -12	+28 +10	0 -8	+16 +4	0 -8	+3 -9	0 -8	-4 -16	0 -8	-8 -20	0 -8	-15 -27
6	10	0 -90	+170 +80	0 -36	+98 +40	0 -15	+35 +13	0 -9	+20 +5	0 -9	+5 -10	0 -9	-4 -19	0 -9	-9 -24	0 -9	-17 -32
10	18	0 -110	+205 +95	0 -43	+120 +50	0 -18	+43 +16	0 -11	+24 +6	0 -11	+6 -12	0 -11	-5 -23	0 -11	-11 -29	0 -11	-21 -39
18	30	0 -130	+240 +110	0 -52	+149 +65	0 -21	+53 +20	0 -13	+28 +7	0 -13	+6 -15	0 -13	-7 -28	0 -13	-14 -35	0 -13	-27 -48
30	40	0 -160	+280 +120	0 -62	+180 +80	0 -25	+64 +25	0 -16	+34 +9	0 -16	+7 -18	0 -16	-8 -33	0 -16	-17 -42	0 -16	-34 -59
40	50	0 -160	+290 +130	0 -62	+180 +80	0 -25	+64 +25	0 -16	+34 +9	0 -16	+7 -18	0 -16	-8 -33	0 -16	-17 -42	0 -16	-34 -59
50	65	0 -190	+330 +140	0 -74	+220 +100	0 -30	+76 +30	0 -19	+40 +10	0 -19	+9 -21	0 -19	-9 -39	0 -19	-21 -51	0 -19	-42 -72
65	80	0 -190	+340 +150	0 -74	+220 +100	0 -30	+76 +30	0 -19	+40 +10	0 -19	+9 -21	0 -19	-9 -39	0 -19	-21 -51	0 -19	-48 -78
80	100	0 -220	+390 +170	0 -87	+260 +120	0 -35	+90 +36	0 -22	+47 +12	0 -22	+10 -25	0 -22	-10 -45	0 -22	-24 -59	0 -22	-58 -93
100	120	0 -220	+400 +180	0 -87	+260 +120	0 -35	+90 +36	0 -22	+47 +12	0 -22	+10 -25	0 -22	-10 -45	0 -22	-24 -59	0 -22	-66 -101
120	140	0 -250	+450 +200	0 -100	+305 +145	0 -40	+106 +43	0 -25	+54 +14	0 -25	+12 -28	0 -25	-12 -52	0 -25	-28 -68	0 -25	-77 -117
140	160	0 -250	+460 +210	0 -100	+305 +145	0 -40	+106 +43	0 -25	+54 +14	0 -25	+12 -28	0 -25	-12 -52	0 -25	-28 -68	0 -25	-85 -125
160	180	0 -250	+480 +230	0 -100	+305 +145	0 -40	+106 +43	0 -25	+54 +14	0 -25	+12 -28	0 -25	-12 -52	0 -25	-28 -68	0 -25	-93 -133
180	200	0 -290	+530 +240	0 -115	+355 +170	0 -46	+122 +50	0 -29	+61 +15	0 -29	+13 -33	0 -29	-14 -60	0 -29	-33 -79	0 -29	-105 -151
200	225	0 -290	+550 +260	0 -115	+355 +170	0 -46	+122 +50	0 -29	+61 +15	0 -29	+13 -33	0 -29	-14 -60	0 -29	-33 -79	0 -29	-113 -159
225	250	0 -290	+570 +280	0 -115	+355 +170	0 -46	+122 +50	0 -29	+61 +15	0 -29	+13 -33	0 -29	-14 -60	0 -29	-33 -79	0 -29	-123 -169

Figure 5.12 *Eight archetypal fits for the 'shaft basis' system of fits*

5.5 Geometry and tolerances

In many instances the geometry associated with tolerances is of significance and the geometry itself needs to be defined by tolerances such that parts fit, locate and align together correctly. Tolerances must therefore also apply to geometric features. The table in Figure 5.13 shows the commonly used geometric tolerance (GT) classes and symbols. These are a selection from ISO 1101:2002. The use of geometric tolerances is shown by three specific examples that are discussed in detail in the following paragraph.

Features and tolerance		Toleranced characteristics	Symbols
Single features	Form tolerances	Straightness	—
		Flatness	⏥
		Circularity	○
		Cylindricity	⌭
Single or related features		Profile of any line	⌒
		Profile of any surface	⌓
Related features	Orientation tolerances	Parallelism	//
		Perpendicularity	⊥
		Angularity	∠
	Location tolerances	Position	⌖
		Concentricity & coaxiality	◎
		Symmetry	⌯
	Run-out tolerances	Circular run-out	↗
		Total runout	⌰

Figure 5.13 *Geometric tolerance classes and symbols*

Figure 5.14 shows the method of tolerancing the centre position of a hole. A 10mm diameter hole is positioned 20mm from a corner. The dimensions show the hole centre is to be 20,00 ± 0,1mm (i.e. a tolerance of ±100um) from each datum face. This means that to pass inspection, the hole centre must be positioned within a 200um square tolerance zone. However, it would be perfectly acceptable for the hole to be at one of the corners of the square tolerance zone, meaning that the actual centre can be 140um from the theoretical centre. This is not what the designer intended and GTs are used to overcome this problem. The method of overcoming this problem is shown in the lower diagram in Figure 5.14. In this case the tolerances associated with the 20mm dimensions are within a GT box. Thus, the 20mm dimensions are only nominal and are enclosed in rectangular squares. The GT box is divided into four compartments. The first compartment contains the GT symbol for position, the next compartment contains the tolerance, and the next two boxes give the datum faces (A and B), being the faces of the corner. Using this GT box, the hole deviation can never be greater than 100um from the centre position.

Figure 5.15 is another example of hole geometry but in this case, the axis of the hole. A dowel is screwed into a threaded hole in a plate. Another plate slides up and down on this dowel. If the axis of the threaded hole is not perpendicular to the top face of the lower plate, the resulting dowel inclination could prevent assembly. By containing the hole axis within a cylinder, the inclination can be limited. The geometrical tolerance box shows the hole axis limits

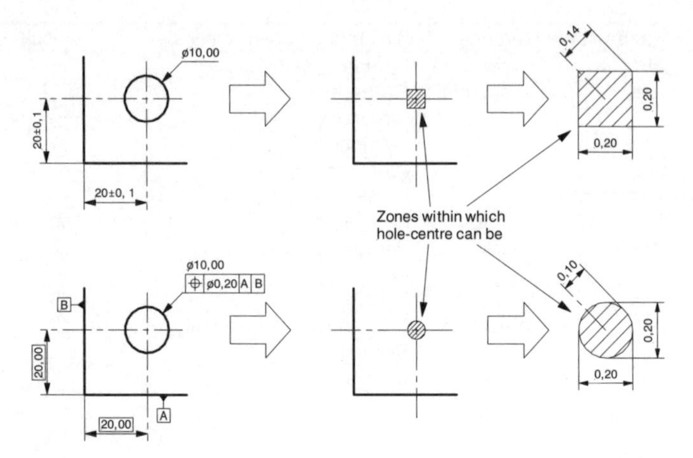

Figure 5.14 *Two methods of tolerancing the centre position of a hole*

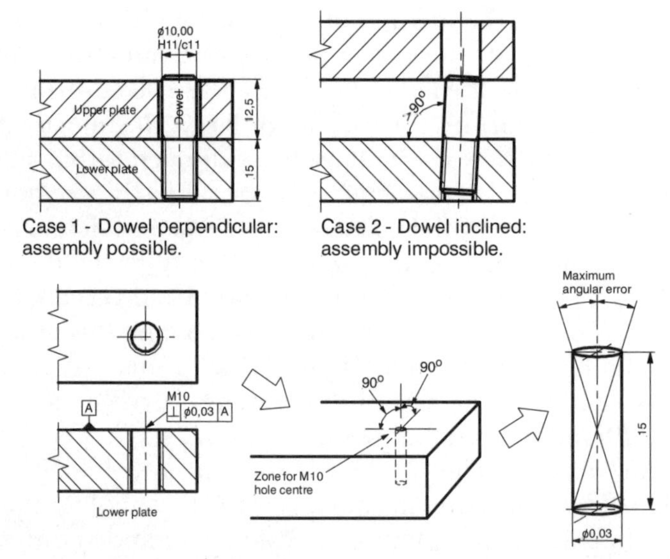

Figure 5.15 *Method of geometric tolerancing the axis perpendicularity of a hole*

which allow assembly. In this case the GT box is divided into three compartments. The left-hand compartment shows the perpendicularity symbol (an inverted 'T') which is shown to apply to the M10 hole, via the leader line and arrow. The right-hand compartment gives the perpendicularity datum that in this case is face 'A'. This is the upper face of the lower plate. This information says that the inclination angle is limited by a cylindrical zone that is 30um in diameter over the length of the hole (the 15mm thickness of the

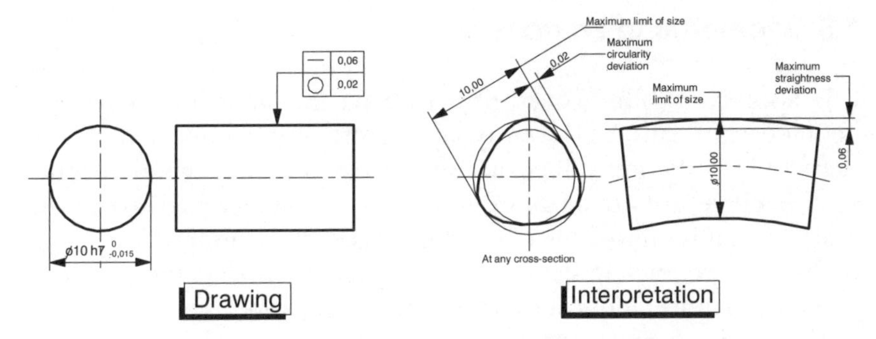

Figure 5.16 *Method of geometric tolerancing straightness and roundness of a cylinder*

lower plate). Thus, the dowel inclination is limited and the upper plate will always assemble.

Figures 5.14 and 5.15 relate to the hole position and axis alignment but nothing has been said about the straightness of the dowel. This situation is considered in the example in Figure 5.16. The dowel has the dual purpose of screwing into the lower plate and locating in the upper plate. If the dowel has a non-circular section or is bent, it may be impossible to assemble. In Figure 5.16, GTs are applied to the outside diameter of the dowel which limits the deviation from a theoretically perfect cylinder. In this case three things are specified using two geometric tolerance boxes and one toleranced feature (the diameter). These are the diametrical deviation, the out-of-roundness and the curvature. The left-hand drawings show the theoretical situation with the cylinder dimensioned in terms of the above three factors. The nominal diameter is 10mm with an h7 tolerance (i.e. 0 and –0,015mm). This means in that whatever position the two-point diameter is measured, the value must be in the range 9,985 to 10,000mm. The out-of-roundness permitted is given in the lower geometric tolerance box. It has two compartments. The left-hand compartment shows the circle symbol (referring to circularity) and the right-hand compartment contains the value of 20um. This means that the out-of-roundness must be contained within two concentric circles that have a maximum circularity deviation of 20um. The upper tolerance box gives the information on straightness. It has two compartments. The left-hand compartment shows the symbol for straightness (a straight line) and the right-hand compartment contains the value 60um. This means that the straightness deviation of any part of the outside diameter outline must be contained within two parallel lines which are separated by 60um.

5.6 Geometric tolerances

GTs apply variability constraints to a particular feature having a geometrical form. A GT can be applied to any feature that can be defined by a theoretically exact shape, e.g. a plane, cylinder, cone, square, circle, sphere or a hexagon. GTs are needed because in the real world, it is impossible to produce an exact theoretical form. GTs define the geometric deviation permitted such that the part can meet the requirements of correct functioning and fit.

Note it is always assumed that if GTs or indeed tolerances in general are not given on a drawing, it is with the assumption that, regardless of the actual situation, a part will normally fit and function satisfactorily.

The chart in Figure 5.13 shows the various geometrical tolerance classes and their symbols given in ISO 1102:2002.

5.6.1 Tolerance boxes, zones and datums

The tolerance box is connected to the feature by a leader line. It touches the box at one end and has an arrow at the other. The arrow touches either the outline of the feature or an extension to the feature being referred to. A tolerance box has at least two compartments. The left compartment contains the GT symbol and the right the tolerance value (see Figure 5.16). If datum information is needed, additional compartments are added to the right. Figure 5.15 shows a three compartment box (one datum) and Figure 5.14 shows a four compartment box (two datums). The method of identifying the datum feature is by a solid triangle which touches the datum or a line projected from it. This is contained in a square box that contains a capital letter. Any capital letter can be used. The datum triangle is placed on the outline of the datum feature referred to or an extension to it.

5.6.2 Geometric tolerance classes

The table in Figure 5.13 has shown the various classes of geometrical tolerance. These are only a selection of the most commonly used ones. The full set is given in ISO 1101:2002.

Row 1 in the table in Figure 5.13 refers to 'GTs of *straightness*'. The symbol for straightness is a small straight line as is seen in the final column of the table. An example of straightness is seen in Figure

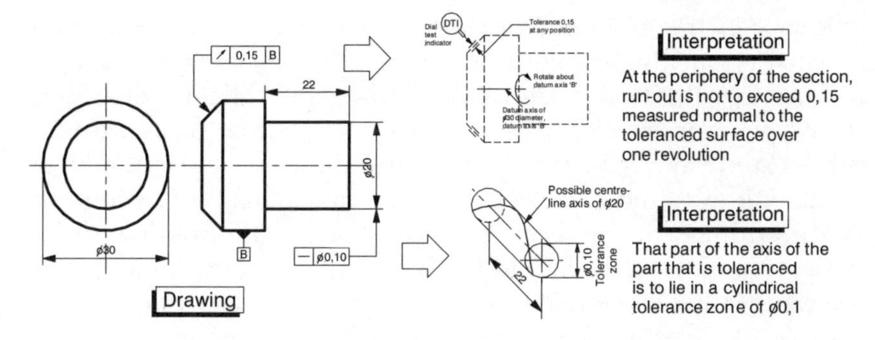

Figure 5.17 *Examples of straightness and runout geometrical tolerancing*

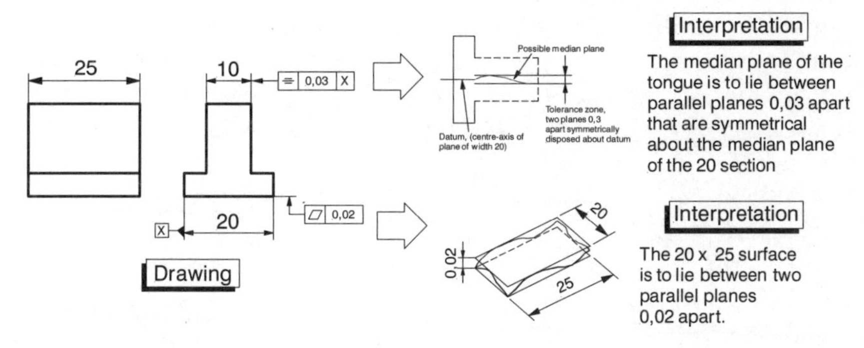

Figure 5.18 *Examples of flatness and symmetry geometrical tolerancing*

5.16. This refers to the straightness of any part of the outline. A straight line rotating about a fixed point generates a cylindrical surface and a GT referring to this is seen in the example of the headed part in Figure 5.17. This is the straightness of the centre axis of the 20mm diameter section. This is the straightness of the axis of a solid of revolution and in this case the tolerance zone is a cylinder whose diameter is the tolerance value, i.e. in this case 100um.

Row 2 in the table in Figure 5.13 refers to GTs of '*flatness*'. The symbol for flatness is a parallelogram. This symbol meant to represent a 3D flat surface viewed at angle. This GT controls the flatness of a surface. Flatness cannot be related to any other feature and hence there is no datum. An example of this is shown in the inverted tee component in Figure 5.18. In this case, the tolerance zone is the space between two parallel planes, the distance between which is the tolerance value. In the case of the example in Figure 5.18, it is the 20um space between the two 20 × 25 mm planes.

Row 3 in the table in Figure 5.13 refers to GTs of '*circularity*'. Circularity can also be called *roundness*. The symbol for circularity is a circle. Circularity GTs control the deviation of the form of a circle in the plane in which it lies. Circularity cannot be related to any other feature and hence there is no need for a datum. For a solid of revolution (a cylinder, cone or sphere) the circularity GT controls the roundness of any cross-section. This is the annular space between two concentric circles lying in the same plane. The tolerance value is the radial separation between the two circles. In the case of the example in Figure 5.16, it is the roundness deviation of the 10mm diameter cylinder given by the 20um annular ring at any cross section.

Row 4 in the table in Figure 5.13 refers to GTs of '*cylindricity*'. The symbol for cylindricity is a circle with two inclined parallel lines touching it on either side. Cylindricity is a combination of roundness, straightness and parallelism. Cylindricity cannot be related to any other feature and hence there is no datum. The cylin-

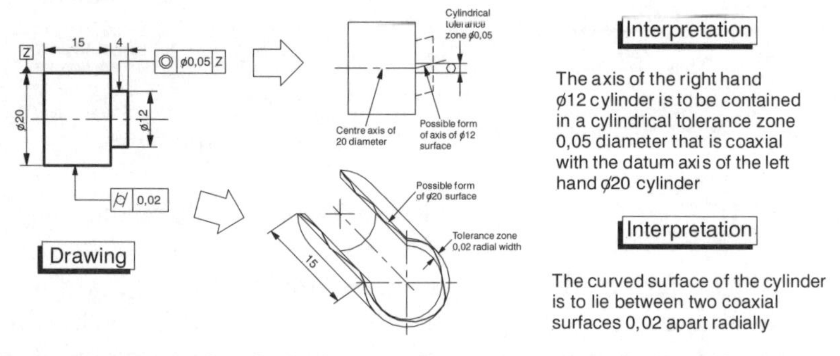

Figure 5.19 *Examples of cylindricity and concentricity geometrical tolerancing*

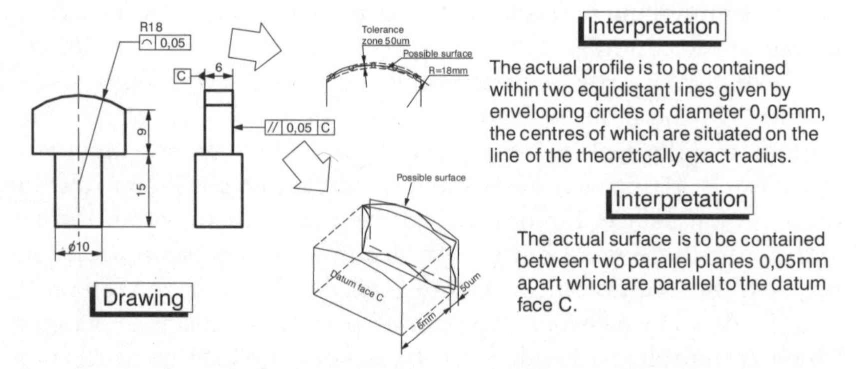

Figure 5.20 *Examples of parallelism and line profile geometrical tolerancing*

dricity tolerance zone is the annular space between two coaxial cylinders and the tolerance value is the radial separation of these cylinders. In the case of the example in Figure 5.19 it is the 20um× 15mm annular cylinder of the 20mm diameter section.

Rows 5 and 6 in the table in Figure 5.13 refer to '*line profile*' and '*area profile*' GTs. The former applies to a line and the latter to an area. The symbol for a line profile GT is an open semicircle and the symbol for an area profile GT is a closed semicircle. These are similar to the straightness (row 1) and flatness (row 2) GTs considered above except that the line and area will be curved in some way or other and defined by some geometric shape. Line or area profiles cannot be related to any other feature and hence there is no datum. The two lines that envelop circles define the line profile tolerance zone. The diameter of these circles is the tolerance value. The centres of the circles are situated on the line having the theoretically exact geometry of the feature. This is to be the case for any section taken parallel to the plane of the projection. An example of a line profile GT is seen in the cam component in Figure 5.20. In this case, the line profile GT means that the profile of any section through the 18mm radius face is to be contained within two equidistant lines given by enveloping circles of 50um diameter about the theoretically exact radius. In the case of area profile GTs, the tolerance zone is limited by two surfaces that envelop spheres. The diameter of these spheres is the same as the tolerance value. The centres of the spheres are situated on the surface having the theoretically exact geometry as the feature referred to.

The remaining rows (7 to 13) in the table in Figure 5.13 are GTs of orientation, location and runout. All these relate to some other feature and hence all require a datum.

Row 7 in the table in Figure 5.13 refers to the first of the GTs that require a datum. These are GTs of '*parallelism*'. The symbol for parallelism is two inclined short parallel lines. The toleranced feature may be a line or a surface and the datum feature may be a line or a plane. In general, the tolerance zone is the area between two parallel lines or the space between two parallel planes. These lines or planes are to be parallel to the datum feature. The tolerance value is the distance between the lines or planes. In the case of the cam in Figure 5.20, the left-hand cam face is to be contained within two planes 50um apart, both of which are parallel to the right-hand face.

Row 8 in the table in Figure 5.13 refers to GTs of '*perpendicularity*'. Perpendicularity is sometimes referred to as *squareness*. The symbol for perpendicularity is an inverted capital 'T'. Note that a perpendicularity GT is a particular case of angularity which is referred to in the next row in the table (row 9). With respect to angularity or squareness, the toleranced feature may be a line, a surface or an axis and the datum feature may be a line or a plane. The tolerance zone is the area between two parallel lines, the space between two planes or, as in the case of Figure 5.15, the space within a cylinder perpendicular to the datum face or plane. In the case of the example in Figure 5.15, the dowel will not assemble with the upper plate if its axis is not within the 30um diameter × 15mm cylindrical tolerance zone which is perpendicular to the upper surface (A) in the lower plate.

Row 9 in the table of Figure 5.13 refers to GTs of '*angularity*'. The symbol for angularity is two short lines that make an angle of approximately 30° with each other. As with perpendicularity, the toleranced feature may be a line, a surface or an axis and the datum feature may be a line or a plane. The tolerance zone is the area between two parallel lines, the space between two planes or the space within a cylinder that is at some defined angle to the datum face or plane. There is no example of angularity in the figures since it is the general case. One could say an example of angularity has been given in Figure 5.15; it just so happens that the angle referred to is 90°.

Row 10 in the table in Figure 5.13 refers to GTs of '*position*'. The symbol for position is a 'target' consisting of a circle with vertical and horizontal lines. The position tolerance zone limits the deviation of the position of a feature from a specified true position. The toleranced feature may be a circle, sphere, cylinder, area or space. In the case of Figure 5.14, it is the position of the centre of the hole with respect to the two datum faces 'A' and 'B' of the corner of the plate.

Row 11 in the table in Figure 5.13 refers to GTs of '*concentricity*'. Concentricity is also referred to as *coaxiality*. The symbol for concentricity is two small concentric circles. This is a particular case of a positional GT (row 10 in the table) in which both the toleranced feature and the datum feature are circles or cylinders. The tolerance zone limits the deviation of the position of the centre axis of a toleranced feature from its true position. An example of this is shown in Figure 5.19. This refers to the concentricity of the smaller 12mm diameter with respect to the larger 20mm diameter section. The centre axis of the 12mm diameter section is to be contained in a cylinder of 50um diameter that is coaxial with the axis of the 20mm diameter section.

Row 12 in the table in Figure 5.13 refers to GTs of '*symmetry*'. The symbol for symmetry is a three bar 'equals' sign in which the middle bar is slightly longer than the other two. A symmetry GT is a particular case of a positional GT in which the position of feature is specified by the symmetrical relationship to a datum feature. In general the tolerance zone is the area between two parallel lines or the space between two parallel planes which are symmetrically disposed about a datum feature. The tolerance value is the distance between the lines or planes. In the case of the tee block in the example in Figure 5.18, the median plane of the 10mm wide tongue is to lie between two parallel lines, 30um apart, which are symmetrically placed about the 20mm wide section of the tee block.

Row 13 in the table in Figure 5.13 refers to GTs of '*runout*'. The symbol for runout is a short arrow inclined at approximately 45°. Runout GTs are applied to the surface of a solid of revolution. Runout is defined by a measurement taken during one rotation of the component about a specified datum axis. A dial test indicator (DTI) contacting the specified surface typically measures it. The tolerance value is the maximum deviation of the DTI reading as it touches the specified surface at any position along its length. An example of a runout GT is shown in the headed shaft in Figure 5.17. In this case, the centre axis of the largest diameter (30mm) is the axis of rotation. The DTI touches the chamfer at any point along its length and, as the component is rotated, the DTI deviation must be within the 150um-tolerance value.

5.7 GTs in real life

When it comes to drawing a part to be manufactured for real, it is not necessary to add GTs to each and every feature. From my experience, the vast majority of features do not need them since the common manufacturing processes achieve the accuracy required in the majority of cases. For example, the dowel perpendicularity in Figure 5.15 is obviously important but provided a sufficiently accurate manufacturing process is chosen, a GT is unnecessary. An understanding of the accuracy that can be achieved by typical manufacturing processes (Figures 4.11 and 5.6) normally negates the need for a GT. However, that having been said, there is usually a need for them to be used where there is a functionally sensitive feature like a shaft running in a journal.

References and further reading

BS 4500A:1985, *Selected ISO Fits, Hole Basis*, 1985.

BS 4500B:1985, *Selected ISO Fits, Shaft Basis*, 1985.

Giesecke F E, Mitchell A, Spencer H C, Hill I L, Dygdon J T, Novak J E and Lockhart S, *Modern Graphics Communications*, Prentice Hall, 1998.

ISO 286–1:2002, *ISO Systems Limits and Fits – Part 1: Basis of Tolerances, Deviations and Fits*, 2002.

ISO 286–2:1988, *ISO Systems of Limits and Fits – Part 2: Tables of Standard Tolerance Grades and Limit Deviations for Holes and Shafts*, 1988.

ISO 1101:2002, *Geometrical Tolerancing – Tolerances of Form, Orientation, Location and Run Out*, 2002.

Mitutoyo, *The Mitutoyo Engineers Reference Book for Measurement & Quality Control*, Mitutoyo (UK) Ltd, 2002.

Zeus Precision Ltd, *Data Charts and Reference Tables for Drawing Office, Toolroom and Workshop*, Zeus Precision Charts Ltd, 2002.

6

Surface Finish Specification

6.0 Introduction

Considering the trace of a supposedly flat surface in Figure 4.11, the 'flat' surface is far from a perfect straight line. Things related to the machine tool, such as vibrations and slide-way inaccuracies cause the long wavelength deviations where the undulations are of the order of millimetres. However, the figure also shows wavelengths of a much smaller magnitude. These deviations are the *surface finish* (SF). They are of the order of tens of microns and they are the machining marks. They are caused by a combination of the tool shape and the feed across the workpiece. In many instances the SF and texture can have a significant influence on functional performance (Griffiths, 2001).

The SF is normally measured by a stylus, which is drawn across the surface to be measured. The stylus moves in a straight line over the surface driven by a traversing unit. This produces a 2D 'line' trace similar to that in Figure 4.11. A line trace produces an X–Y set of data points that can be analysed in a variety of statistical ways to produce parameters. These parameters are descriptors of a surface. They can be used to describe the SF of a surface in much the same way as a dimension describes the form of a feature. In the same way that a dimension can never be exact, the SF, represented by a parameter, can never be exact. Tolerances also need to added to SF specifications. To ensure fitness for purpose, the SF needs to be defined with limits. This chapter is concerned with the specification of SF and texture.

6.1 Roughness and waviness

A trace across a surface provides a profile of that surface which will contain short and long wavelengths (see Figure 4.11). In order for a surface to be correctly inspected, the short and long wavelength components need to be separated so they can be individually analysed. The long waves are to do with dimensions and the short waves are to do with the SF. Both can be relevant to function but in different ways. Consider the block in Figure 6.1. This has been produced on a shaping machine. The block surface undulates in a variety of ways. There is a basic roughness, created by the tool feed marks, which is superimposed on the general plane of the surface. Thus, one can identify two different wavelengths, one of a small scale and one of a large scale. These are referred to as the *roughness* and *waviness* components.

Roughness and waviness have different influences on functional performance. A good example illustrating the differences concerns automotive bodies. Considering the small-scale amplitudes and wavelengths called 'roughness', it is the roughness, not waviness, which influences friction, lubrication, wear and galling, etc. The next scale up from roughness is 'waviness' and it is known that the visual appearance of painted car bodies correlates more with waviness than roughness. The reason for this is the paint depth is about 100um and it has a significant filtering effect on roughness but not waviness.

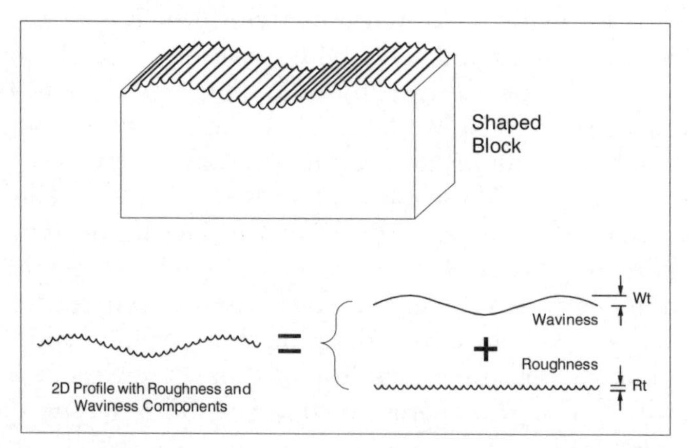

Figure 6.1 *A shaped block showing roughness and waviness components*

6.2 Measuring the surface finish

The most common method of assessing the SF is by traversing a stylus across a surface. A typical stylus is shown in the scanning electron microscope (SEM) photograph in Figure 6.2 (courtesy of Hommelwerke GmbH). The stylus tip is made of diamond having a tip spherical radius of 5um and an included cone angle of 90°. Styli are available in a standard range of spherical radii of 2, 5 and 10um and included cone angles of 60° and 90° (ISO 3274:1996). The stylus is shown in contact with a ground surface that gives an indication of the scale of the surface features. The stylus is positioned at the end of a mechanical arm that connects to a transducer such that the undulations on the surface are translated into an electrical signal. This signal is amplified and eventually displayed on a PC screen along with the calculated parameters.

6.2.1 Sample length and evaluation length

Considering the case of a flat surface, the traverse unit drives the stylus over a distance called the *evaluation length* (EL). This length is

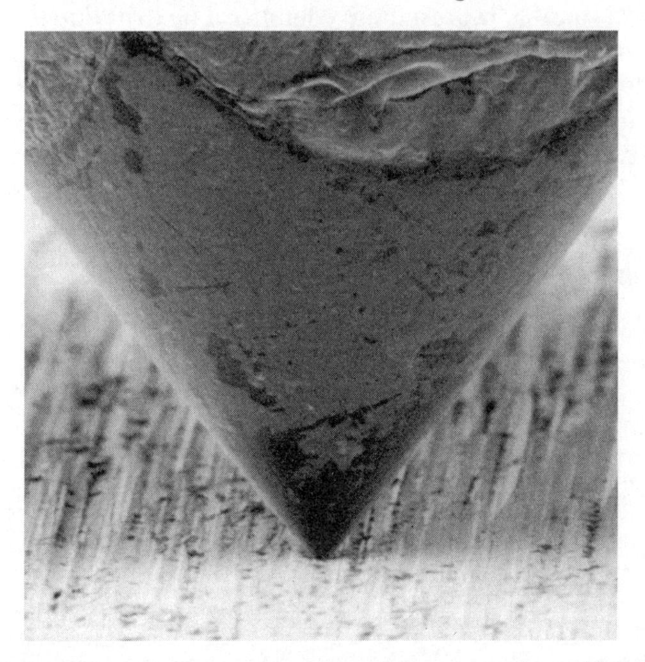

Figure 6.2 *A scanning electron microscope photograph of a stylus (courtesy of Hommelwerke GmbH)*

divided into five equal parts, each of which is called a *sampling length* (SL). In ISO 4287:1997, the sample length is defined as the '*length in the direction of the X-axis used for identifying the irregularities characterising the profile under evaluation*'. The evaluation length is defined as the '*length in the direction of the X-axis used for assessing the profile under evaluation*'.

The SL length is significant and is selected depending upon the length over which the parameter to be measured has statistical significance without being long enough to include irrelevant details. This limit will be the difference between roughness and waviness. In Figure 6.3, the waviness is represented by the sine wave caused by such things as guideway distortion. The roughness is represented by the cusp form caused by the tool shape and micro-roughness by the vees between cusps caused by tearing. The SL over which the profile is assessed is critical, if it is too large (L1) then waviness will distort the picture, if it is too small (L2) then the unrepresentative micro-roughness will only be seen. The correct SL is that length over which the parameter to be measured is significant without being so long as to contain unwanted and irrelevant information. The length (L3), containing several feed-rate cycles, would be a suitable representative length. The drift due to the wavelength would be filtered out.

The default SL is 0,8mm. This is satisfactory for the vast majority of situations but for processes that use a very small or a very large feed, this is inappropriate. Information on how to determine the correct SL for non-standard situations is given in ISO 4288:1996.

6.2.2 Filters

A filter is a means of separating roughness from waviness. Mummery (1990) gives the useful analogy of a garden sieve. A sieve

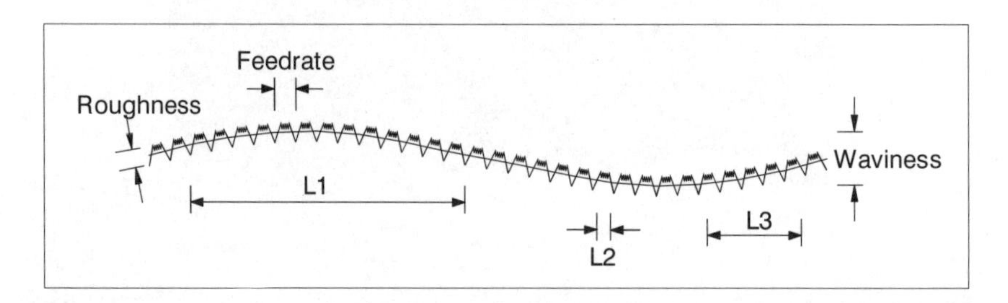

Figure 6.3 *The effect of different sampling lengths*

separates earth into two piles. One could be called rock and the other dirt. The sieve size and therefore the distinction between dirt and rock is subjective. A gardener would use a different sieve size in comparison to a construction worker. With reference to machine surfaces, a sieve hole size is analogous to the filter. Figure 6.4 shows the results of different types of filters.

The simplest filter is the *2CR filter*. It consists of two capacitors and two resistors. With the 2CR filter, there is 75% transmission for a profile with a 0.8mm wavelength. This is because all filter design is a compromise; 100% transmission up to the cut-off value and nothing after is impractical. In practice, the 2CR filter produces a phase shift and overshoot because it cannot read ahead. The 2CR filter is not mentioned in the latest standards.

The *phase corrected (PC) filter* (ISO 11562:1996) overcomes some of the disadvantages of the 2CR filter in that it can look forward. It does this by the use of a window or mask similar to that used in digital image processing. The mask or window of a PC filter is called a *weighted function*. The mask is 1D and consists of a series of weights arranged in a Gaussian distribution. Each weight is applied to each profile point over the length of the window. Shifting the mask step by step scans the profile.

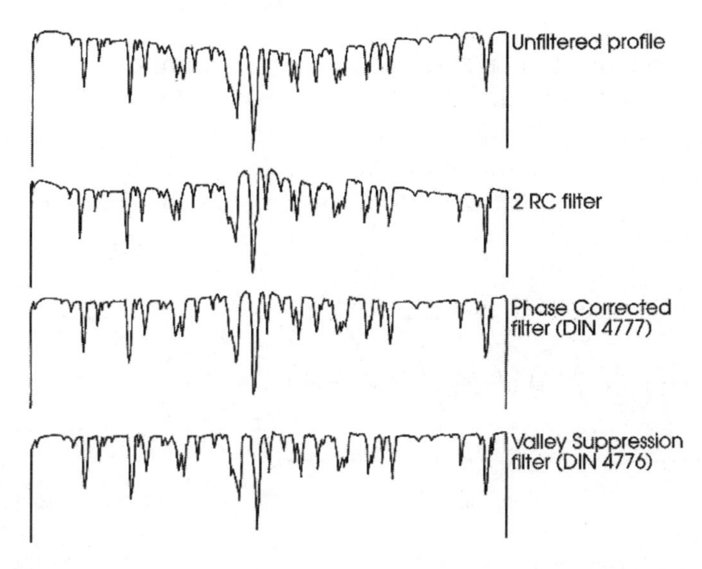

Figure 6.4 *The effect of 2CR, phase corrected (PC) and valley suppression (VS) filters on a profile*

The PC filter will still produce errors particularly with the highly asymmetric profiles. For example, deep valleys will cause a distortion because of their comparative 'weight' within the mask. To overcome the above disadvantage, a double filter is applied which has the effect of suppressing valleys even further. This is called the valley suppression (VS) filter or the double Gaussian filter. It is defined in ISO 13565–1:1996.

Figure 6.4 (Mummery, 1990) shows a comparison of the 2CR, PC and VS filters when applied to a plateau-honed surface. The 2RC filter produces a 'bump' distortion in the region of the centre-left deep valley. This distortion is reduced but not eliminated by the PC filter in that a slight raising of the profile can still be seen at the same centre-left valley. The double filter reduces this to an almost negligible amount.

6.3 Surface finish characterization

Once a satisfactory profile is obtained, it can be analysed and represented by a variety of means. This raises the question of what particular number, parameter or descriptor should be used. Unfortunately, there is no such thing as a universal parameter or descriptor and one must select from the ones published in the ISO standards.

With reference to Figure 6.5, the ADF (Amplitude Distribution Function or height distribution function) is a histogram where the value of p(y) represents the fraction of heights lying in the stratum between y and (y + dy). If the ADF is integrated, the BAC or *Abbott-Firestone Curve* or *Material Ratio Curve* is obtained. The BAC can also

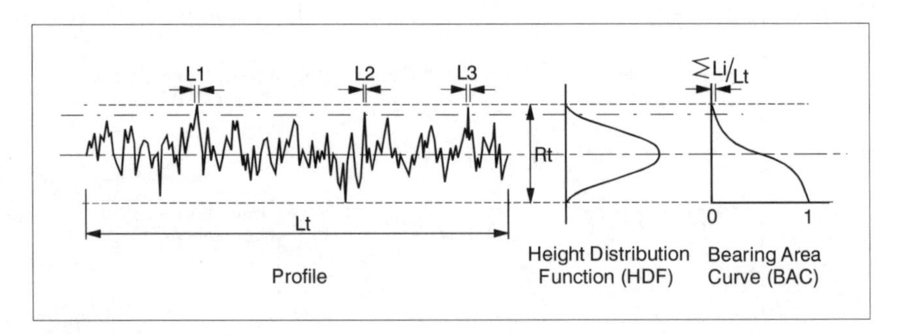

Figure 6.5 *A profile and the corresponding height distribution function and bearing area curve*

be generated by slicing the profile in a straight line parallel to the mean from the highest peak down, plotting the total length revealed as a fraction of the profile length under consideration. This is the equivalent of a perfect abrasion or wear process. Examples of the graphical outputs as well as parameters are shown in Figure 6.6. This is a trace from a fine-turned surface, showing the conventional turning unit event 'cusp' surface form. The peak spacing is approximately 115um and the peak to valley height is 45um.

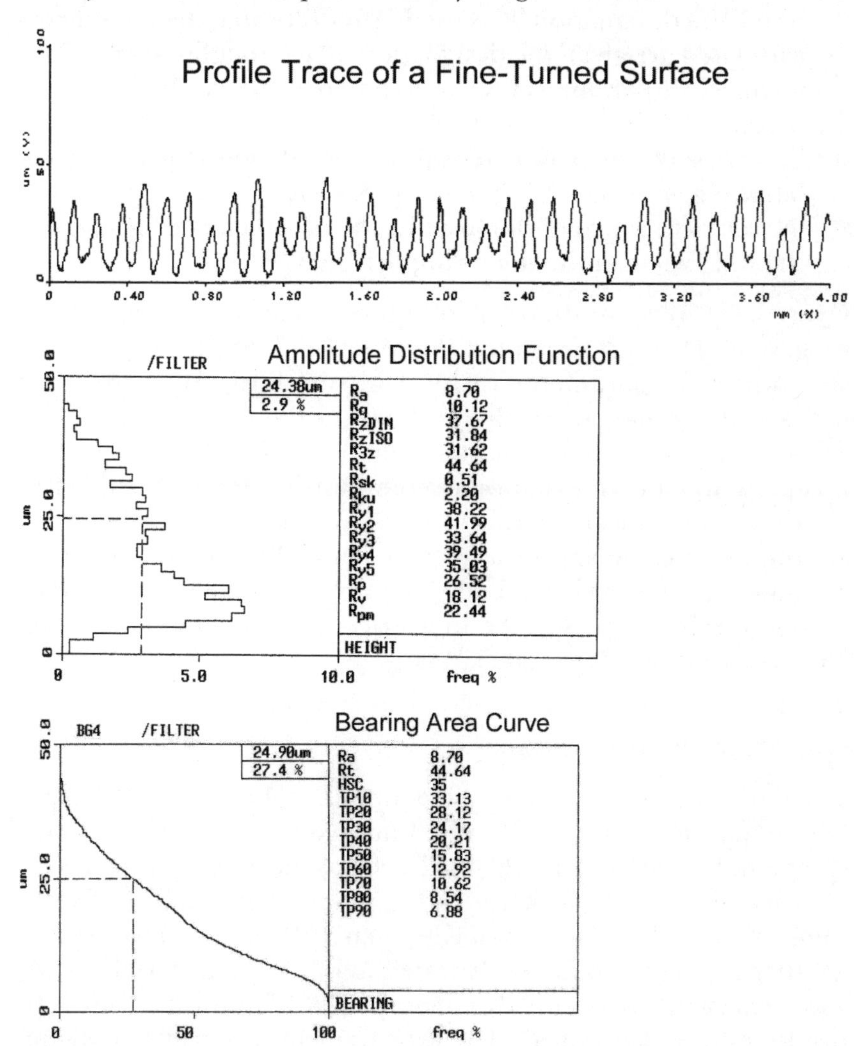

Figure 6.6 *A profile of a fine-turned surface and the corresponding ADF and BAC curves*

6.3.1 2D roughness parameters

The range of parameters calculated from a trace may be represented by the equation:

$$\text{parameter} = \text{TnN}$$

where:

- 'T' represents the scale of the parameter. If the trace is unfiltered, the designation 'P' is used. After filtering, the parameters calculated are given the designation 'R' for roughness or 'W' for waviness. If parameters relate to an area, the designation 'S' is used.
- 'n' represents the parameter suffix which denotes the type calculated, e.g. *average* is 'a', *RMS* is 'q', *Skew* is 'sk', etc.
- 'N' refers to which of the five SLs the parameter relates to, e.g. the RMS value of the third sample is Rq3.

Over the years, hundreds of roughness parameters have been suggested. This has prompted Whitehouse (1982) to describe the situation as a 'parameter rash'! The standard ISO 4287:1997 defines 13 parameters which are shown in the table in Figure 6.7. These parameters are the most commonly used ones and the ones accepted by the international community as being the most relevant. They are divided into classes of heights, height distribution, spacing and angle (or hybrid). It should be noted that there are other parameters, based on shapes of peaks and valleys, which are more relevant to specific industries like the automotive (ISO 13565–2:1996 and ISO 12085:1996).

6.3.1.1 2D amplitude parameters

The table in Figure 6.8 gives the definitions of the ISO 4287:1997 height parameters. The centre line average (Ra) is the most common. It is defined in ISO 4287:2000 as the '*arithmetic mean deviation of the assessed profile*'. Over an EL, there will normally be five Ra values, Ra1 to Ra5. The root mean square (RMS) parameter (Rq) is another average parameter. It is defined in ISO 4287:1997 as the '*root mean square deviation of the assessed profile*'. There will normally be five Rq values: Rq1 to Rq5. The Rq parameter is statistically significant because it is the standard deviation of the profile about the mean line.

PARAMETER CLASS	PARAMETERS IN ISO 4287
Heights	Ra, Rq, Rv, Rp, Rt, Rz, Rc
Height Distribution	Rsk, Rku, Rmr, Rmr(c)
Spacing	Rsm
Hybrid	RΔq

Figure 6.7 *The 2D roughness parameters given in ISO 4287:2000*

With respect to parameters which measure extremes rather than averages, the Rt parameter is the value of the vertical distance from the highest peak to lowest valley within the EL (see Figures 6.8 and 6.9). It is defined in ISO 4287:1997 as the '*total height of profile*'. There will be only one Rt value and this is THE extreme parameter. It is highly susceptible to any disturbances. The maximum peak to valley height within each SL is Rz (see Figures 6.8 and 6.9). It is defined in ISO 4287:1997 as the '*maximum height of the profile*'. There are normally five Rz values, Rz1 to Rz5, or Rzi. With reference to the fine-turned profile of Figure 6.6, the Rzi values are shown as Ryi, a former designation.

Material above and below the mean line can be represented by peak and by valley parameters (see Figures 6.8 and 6.9). The peak parameter (Rp) is the vertical distance from the highest peak to the

PROFILE HEIGHT PARAMETERS		
Parameter		Description
Ra	Centre Line Average	$Ra = \frac{1}{n}\sum_{i=1}^{n}\lvert y_i\rvert = \frac{1}{L_r}\int_0^{L_r}\lvert y\rvert\,dx$
Rq	RMS Average	$Rq = \sqrt{\frac{1}{n}\sum_{i=1}^{n} y_i^2} = \sqrt{\frac{1}{L_r}\int_0^{L_r} y^2\,dx}$
Rt	EL peak to valley height	Peak to valley height within the EL
Rz	SL peak to valley height	Peak to valley height within a SL
Rp	Peak height	Highest peak to mean line height
Rv	Valley depth	Lowest valley to mean line depth

Figure 6.8 *The 2D height parameters given in ISO 4287:2000*

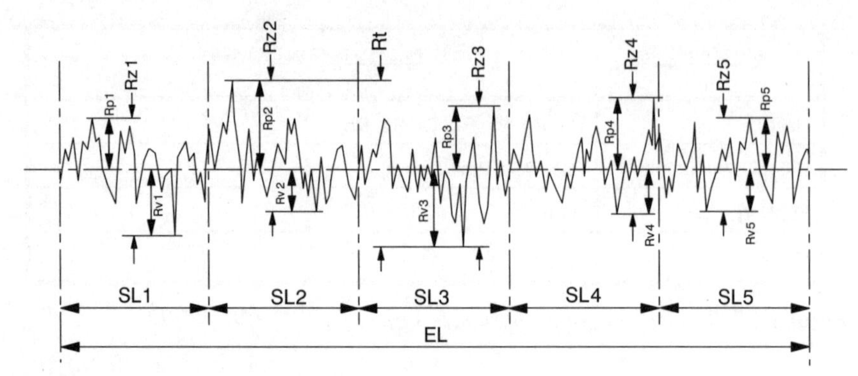

Figure 6.9 *A schematic profile and the parameters Rt, Rz, Rv, Rp*

mean line within a SL. It is defined in ISO 4287:1997 as the '*maximum profile peak height*'. The valley parameter, Rv, is the maximum vertical distance between the deepest valley and the mean line in a SL. It is defined in ISO 4287:1997 as the '*maximum profile valley depth*'.

6.3.1.2 2D amplitude distribution parameters

With respect to a profile, the sum of the section profile lengths at a depth 'c' measured from the highest peak is the material length (Ml(c)). In ISO 4287:1997 the parameter Ml(c) is defined as the '*sum of the section lengths obtained by a line parallel to the axis at a given level, "c"*'. This is the summation of 'Li' in Figure 6.5. If this length is expressed as a percentage or fraction of the profile, it is called the '*material ratio*' (Rmr(c)) (see Figure 6.10). It is defined in ISO 4287:1997 as the '*ratio of the material length of the profile elements Ml(c) at the given level "c" to the evaluation length*'. In a previous standard, this Rmr(c) parameter is designated 'tp' and can be seen as TP10 to TP90 in the fine-turned BAC of Figure 6.6.

The shape and form of the ADF can be represented by the function moments (m_N):

$$m_N = \frac{1}{n}\sum_{i=1}^{n} y_i^N \qquad = \frac{1}{L}\int_0^L y^N \, dx$$

where N is the moment number, y_i is the ordinate height and 'n' is the number of ordinates. The first moment (m_1) is zero by definition. The second moment (m_2) is the variance or the square of the

PROFILE HEIGHT DISTRIBUTION PARAMETERS		
Parameter		Description
Rmr(c)	Material ratio at depth 'c'	$Rmr(c) = \frac{1}{Ln}\sum_{i=1}^{n} L_i = \frac{Ml(c)}{Ln}$
Rsk	Skew	$Rsk = \frac{1}{Rq^3}\left[\frac{1}{n}\sum_{i=1}^{n} y_i^3\right] = \frac{1}{Rq^3}\left[\frac{1}{Lr}\int_0^{Lr} y^3 dx\right]$
Rku	Kurtosis	$Rku = \frac{1}{Rq^4}\left[\frac{1}{n}\sum_{i=1}^{n} y_i^4\right] = \frac{1}{Rq^4}\left[\frac{1}{Lr}\int_0^{Lr} y^4 dx\right]$

Figure 6.10 *The 2D height distribution parameters given in ISO 4287:2000*

standard deviation, i.e. Rq. The third moment (m_3) is the skew of the ADF. It is usually normalised by the standard deviation and, when related to the SL, is termed Rsk. It is defined in ISO 4287:1997 as the '*skewness of the assessed profile*'. For a random surface profile, the skew will be zero because the heights are symmetrically distributed about the mean line. The skew of the ADF discriminates between different manufacturing processes. Processes such as grinding, honing and milling produce negatively skewed surfaces because of the shape of the unit event/s. Processes like sandblasting, EDM and turning produce positive skewed surfaces. This is seen in the fine-turned profile in Figure 6.6 where the Rsk value is +0.51. Processes like plateau honing and gun-drilling produce surfaces that have good bearing properties, thus, it is of no surprise that they have negative skew values. Positive skew is an indication of a good gripping or locking surface.

The fourth moment (m_4) of the ADF is *kurtosis*. Like the skew parameter, kurtosis is normalised. It is defined in ISO 4287:1997 as the '*kurtosis of the assessed profile*'. In this normalised form, the kurtosis of a Gaussian profile is 3. If the profile is congregated near the mean with the occasional high peak or deep valley it has a kurtosis greater than 3. If the profile is congregated at the extremes it is less than 3. A theoretical square wave has a kurtosis of unity.

6.3.1.3 2D spacing parameters

Figure 6.11 shows a schematic profile of part of a surface that has been turned at a feed of 0,1mm/rev. The cusp profile is modified by small grooves caused by wear on the tool. The problem with this profile is that there are 'macro' and 'micro' peaks, the former being at 0,1mm spacing and the latter at 0,011mm spacing. Either could be important in a functional performance situation. This begs the question, 'when is peak a peak a peak?' To cope with the variety of possible situations, many spacing parameters have been suggested over the years. However, it is unfortunate that in the ISO standard only one parameter is given. This is the average peak spacing parameter RSm that is the spacing between peaks over the SL at the mean line. It is defined in ISO 4287:1997 as the '*mean value of the profile element widths within a sampling length*'. With respect to Figure 6.11, if the 0,2mm were the SL, there are 10 peaks shown and hence RSm = 0,02mm.

6.3.1.4 2D slope parameters

The RMS average parameter (RΔq) is the only slope parameter included in the ISO 4287:1997 standard. It is defined as the '*root mean square of the ordinate slopes dz/dx within the sampling length*'. There will normally be five RΔq values for each of the SL values: RΔq1 to RΔq5. The RΔq value is statistically significant because it is the standard deviation of the slope profile about the mean line. Furthermore, the slope variance is the second moment of the slope distribution function. In theory, there can be as many slope parameters as there are height parameters because parameters can be just as easily be calculated from the differentiated profile as from the original profile.

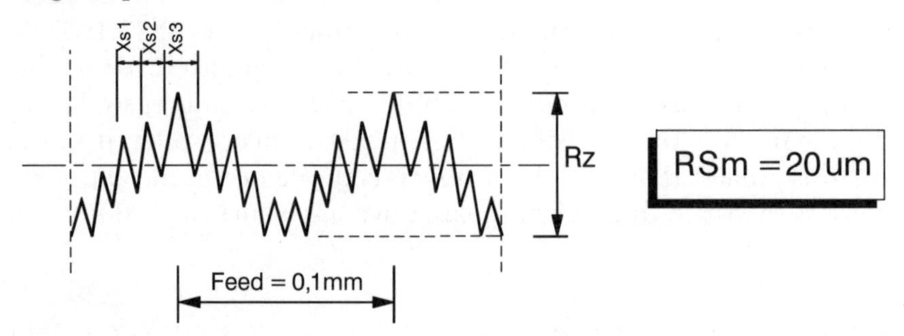

Figure 6.11 *The 2D spacing parameter given in ISO 4287:2000*

6.4 Tolerances applied to the assessment of surface finish

The SL sets the limits for the horizontal length to be considered along the surface. By definition, there also needs to be limits defined in the other direction (the vertical). This defines the deviation allowed perpendicular to the surface. This will be the SF tolerance. Like any length dimension, the SF tolerance needs to be in the form of a tolerance band or range within which the 2D parameter may vary. There are two types of tolerance. Firstly, there is an upper one that the measured value must not be greater than and secondly, an upper one and a lower one that the measured value must not be less than. In the first case, there is only one value and this is the upper one. No lower one is specified but, in the case of, say, height parameters it is effectively zero because this is the lowest practical limit.

The standard ISO 4288:1996 provides flexibility with respect to the acceptance or rejection of the measured surface when compared with a tolerance because there are two rules specified in the standard, the '*16%-rule*' and the '*max-rule*'. The '16%-rule' allows some of the values to be greater than the upper limit or less than the lower limit (see Figure 6.12). With respect to the upper limit, the surface is considered acceptable if not more than 16% of the measured values of the selected parameter exceed the value specified on an engineering drawing. With respect to the lower limit, the

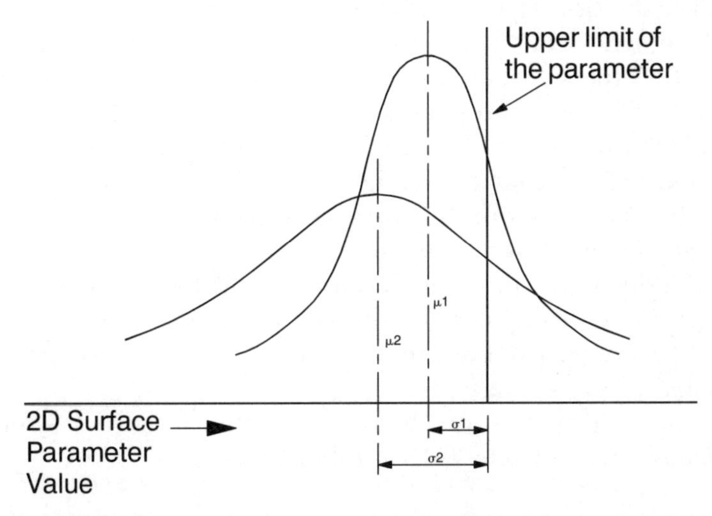

Figure 6.12 *The 16%-rule and the upper limit for two distributions (ISO 4288:1996)*

surface is considered acceptable if not more than 16% of the measured values of the selected parameter are less than the value specified. In cases where the surface parameter being inspected follows a normal distribution, the 16%-rule means that the upper limit is located at a value of the $\mu + \sigma$ where μ is the mean value and σ is the standard deviation of the values. The greater the value of the standard deviation, the further from the specified limit the mean value of the roughness parameter needs to be.

6.5 Method of indicating surface finish and texture

Section 6.3.1 above described parameters using '*TnN*'. However, no information was given concerning how these are added to features on a drawing. The methodology to do this is described in ISO 1302:2001. It is based on what is termed a 'tick symbol' that defines the SF and points to the surface in question via a leader line.

Figure 6.13 shows the tick symbol with various descriptors surrounding it. The tick symbol is placed on the surface or an extension drawn to it. The basic tick comprises two lines at 60° to each other. This basic open tick (Figure 6.13a) has no significance of its own. Closing the tick symbol (Figure 6.13b) indicates that the surface must be machined. If machining is prohibited for some reason, for example, residual stresses must not be added, a circle is placed over the tick (Figure 6.13c). When additional information is to be added, a horizontal line is added to the right tick arm (Figure 6.13d). When the same surface texture is required on all surfaces around a workpiece, represented on an orthographic 2D drawing by a closed outline, a circle is added to the symbol at the junction of the tick and the horizontal line (Figure 6.13e). It is the symbol that means 'all surfaces around a workpiece outline'. For example, consider the gauge shown in Figure 6.14. A surface roughness 'tick' symbol is added to the top face. Because the tick has the small circle on it, the surface roughness requirement applies the eight faces around the front view but not the front face, shown as face (a), nor the back face, shown as face (b).

Additional information can be added to the closed tick symbol and arm as shown in Figure 6.15 as follows:

- ***Positions*** '*a, b and c*' – the surface texture parameters, numerical values, transmission band and SL information are placed at

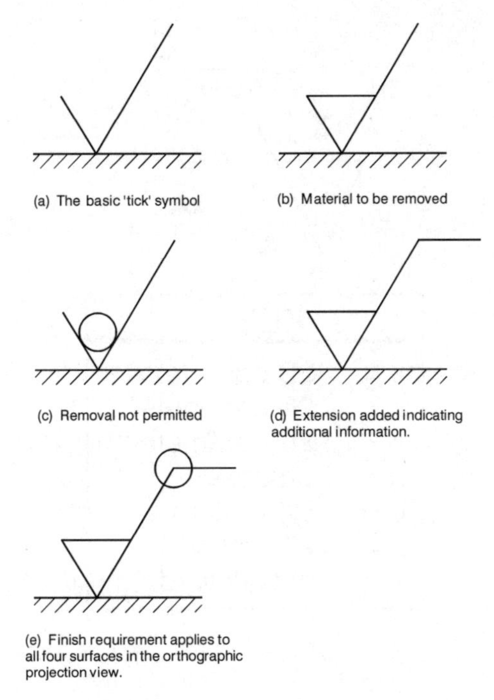

Figure 6.13 *The 'tick' symbol of ISO 1302:2001*

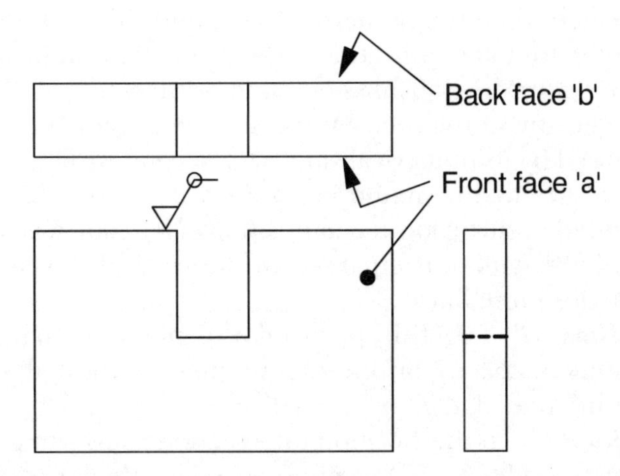

Figure 6.14 *A component that has the same surface finish requirement on 8 of its 10 faces*

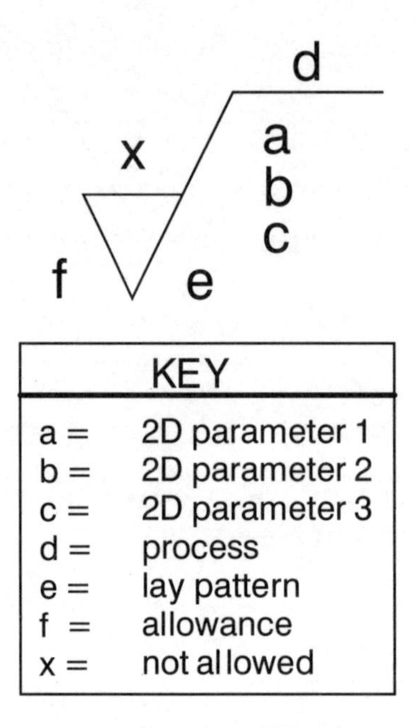

Figure 6.15 *The position of additional information to be added to the 'tick' symbol*

positions a, b and c. If only one single SF parameter is to be specified, then the numerical value and the transmission band and SL are to be indicated in the complete graphical symbol at position a. The transmission band or SL is to be followed by an oblique stroke followed by the surface parameter designation, followed by its numerical value. If a second surface parameter is to be specified it should be located at position b. If a third is required it will be located at position c. If a fourth is required the graphical symbol is enlarged in the vertical direction to make room for more lines.

- ***Position** 'd'* – at this position the manufacturing method, treatment, coating or other requirement is located, e.g. turned, ground, plated, etc.
- ***Position** 'e'* – at this position information concerning the lay and orientation is given. A symbol represents the lay pattern. There are seven lay classes represented by the symbols: '=, ⊥, X, M, C, R and P'. These are shown in the table in Figure 6.16.
- ***Position** 'f'* – at this position the required machining allowance is indicated as a numerical value in millimetres. The machining

allowance is generally indicated only in those cases where more than one processing stage is shown on one drawing. Machining allowances are therefore found, for example, in drawings of raw, cast or forged workpieces.

- ***Position*** '*x*' – no SF indications are to be added above the tick symbol at position x. This may seem a peculiar thing to say but in previous standards, only the Ra value was to be placed at this position, all other parameters were to be placed at a different position. This implied that the Ra value had a prominence over other parameters and that it was the most important parameter of all.

The full designation attached to positions a, b and c of a tick symbol could contain up to seven elements. Consider the following as an example:

U 'X'0,08–0,8/Rz2max 3,3

Graphical Symbol	Interpretation	Lay Pattern
=	Parallel to the plane of projection of the view in which the symbol is used.	
‖	Perpendicular to the plane of projection of the view in which the symbol is used.	
X	Crossed in two oblique directions relative to the plane of projection of the view in which the symbol is used.	
M	Multi-directional.	
C	Approximately circular relative to the centre of the surface to which the symbol applies.	
R	Approximately radial relative to the centre of the surface to which the symbol applies.	
P	Lay is particulate, non-directional or protuberant.	

Figure 6.16 *Symbols for surface lay according to ISO 1302:2001*

The interpretation of this is as follows. The first specification, the 'U', means the upper specification limit that applies to the parameter Rz in the second SL (Rz2). In this instant there is no lower value and the Rz parameter in theory could be 0. If there is a lower limit then the capital letter 'L' is shown. If neither 'U' nor 'L' is shown, it is assumed to be the upper limit (U). The second specification, shown as 'X' in the above is the filter (see Section 6.2.2 above). It should be noted that a range of something like 12 filter standards will be published as Technical Specifications (ISO 16610) in 2002 and 2003. The third and fourth specifications are the transmission band limits, shown in this case as 0,08–0,8. These are the short wave and long wave filters. The fifth specification is the 2D parameter itself, in this case the Rz value in the second SL. The sixth specification in the above is the '16% rule' or the 'max-rule', in this case the 'max-rule'. The seventh specification is the parameter limit value, in this case 3,3um (Rz2).

Figure 6.17 shows examples of the use of the tick symbol. The interpretation shown in Figure 6.17a is as follows. The process is not specified therefore any which meets the roughness specification is acceptable. The parameters specified apply to the roughness

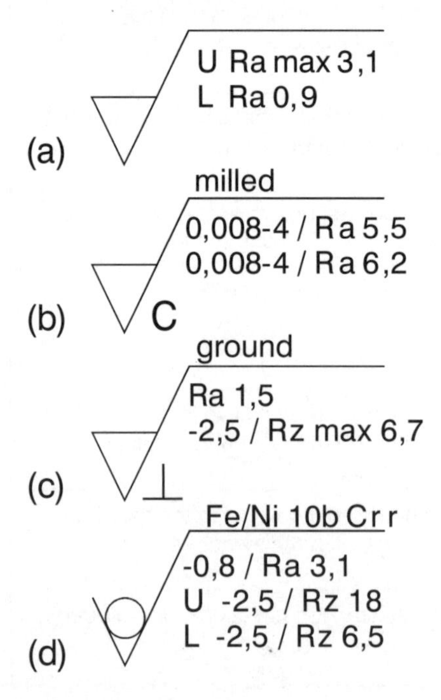

Figure 6.17 *Examples of tick symbol designations*

profile. The upper limit is a Ra value of 3,1um using the 'max-rule'. The lower limit is a Ra value of 0,9um and the '16% rule' applies as the default. With both the upper and lower limits, the default transmission bands apply and the Ra value is to be examined over the EL of five SL's. The lay is unspecified so therefore any lay pattern is acceptable.

The interpretation of the tick symbol shown in Figure 6.17b is as follows. The upper specification limit is a Ra value of 5,5um and the lower specification limit is a Ra value of 6,2um. The '16% rule' applies to both as the default. In both cases the lower transmission band is 0,008mm and the upper is 4mm. No SL is specified with respect to the parameter so the default situation of the EL applies. Thus, all of the five SL Ra values must be tested. The manufacturing processes is to be milling such that the surface layer is approximately circular around the centre. In this case the 'U' and 'L' is not stated because they are obvious.

The interpretation of the tick symbol as shown in Figure 6.17c is as follows. The surface is to be produced by grinding with the lay approximately perpendicular to the projection plane. There are two upper specification limits set by a Ra value and a Rz value. The Ra value is limited to 1,5um using the default '16% rule'. The upper and lower transmission bands are the default values. All five SL's are to be considered. The Rz value is to be limited to 6,7um and the 'max-rule' applies. The upper transmission band is 2,5mm and the lower transmission band is the default value. All five Rz values in an EL must be considered.

The interpretation of the tick symbol in Figure 6.17d is as follows. The surface is to be produced by nickel-chromium plating and no material is to be removed afterwards. There is one upper limit for Ra and an upper and a lower limit for Rz. The upper limit for Ra is 3,1um and '16% rule' applies. The lower transmission band is the default value and the upper transmission band is 0,8mm. Each of the five SL's is to be examined for the Ra values. The upper limit is 18um when the lower transmission band is the default value and the upper transmission band is 2,5mm. The lower limit is Rz of 6,5um when the lower transmission band is the default value and the upper is 2,5mm. The '16% rule' applies to the both the upper and the lower Rz values. All five SL's are to be investigated with respect to the upper and lower Rz values.

6.6 3D surface characterization

At present there is no 3D parameter standard. It is too early in the development cycle. Research is still needed to explore the possibilities and provide recommendations. A research programme undertaken by Birmingham University has led to a proposal for some 3D parameters and that they should have the parameter designation 'S' for 'surface' (Stout et al, 1993 and 2000). The proposal was for the recognition of a primary set of 14 3D parameters. They are mostly 3D versions of 2D parameters, e.g. Rq to Sq. However, proposals were made concerning areal bearing area parameters which seem particularly useful.

Research continues at various establishments and two EU funded research programmes are of note that were reported in 2001. One was concerned with 3D parameter specification in general (called *Surfstand*) and the other was concerned with the 3D assessment of automotive body panels (called *Autosurf*). These two reports recommend that a series of 3D parameters should be defined in two Technical Reports that will be published in 2002 as consultation documents. Since these 3D parameters are at present more appropriate to the laboratory than the factory, they will not be discussed any further here.

6.7 Surface finish specification in the real world

When it comes to drawing a part to be manufactured for real, it is not necessary to add an SF specification to each and every feature. The vast majority of features do not need them since the common manufacturing processes achieve the SF required and more often than not, the SF is unimportant. It is only in a few instances, where a surface is functionally important, that it is necessary to define a SF. Indeed, specifying a SF is the exception rather than the rule and I have seen many drawings that do not have any SF specifications on them at all!

Note that the vice assembly drawing in Figure 3.1 has no SF specification. This should not be surprising since it is an assembly drawing with no manufacturing information. The movable jaw drawing in Figure 3.2 has just one SF specification. This is for the two bottom surfaces of the jaw where it contacts the body. In this case a fine SF ($Rz < 0{,}2um$) is required to minimise friction and ease movement. Such a fine SF can be easily achieved by polishing. Although not shown, there would be a complementary SF specification on the body detail drawing. There are no SF requirements on the hardened insert drawing in Figure 3.3 simply because they are not needed for the correct functioning of the vice.

With regard to the SF parameter values produced by common manufacturing processes, it is unfortunate that few SF parameter values have been published but many have been published in research papers. Books that give details of some SF parameter values are those of Dagnall (1998) and Mummery (1990). Griffiths (2001) gives the results of an investigation linking 2D and 3D SF parameters to common manufacturing processes. The graph in Figure 6.18 compares surface heights and lengths in the form of the 2D parameters Rz and RSm. These parameters are the average height and average length and therefore represent the average 'unit event' dimensions. The ratio of length to height varies from less than 10:1 to greater than 100:1, with an average in the region of

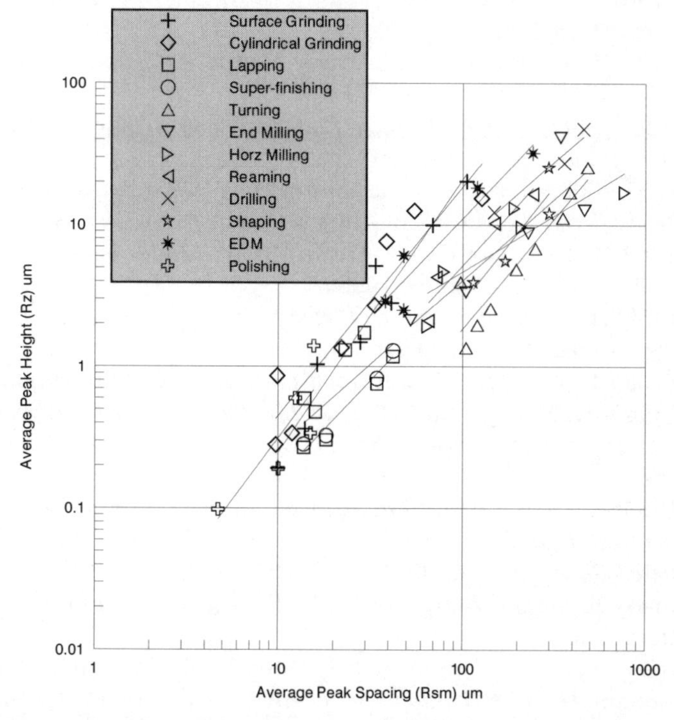

Figure 6.18 *Surface finish parameters Rz and RSm for a range of common manufacturing processes*

10:1. On the diagram, best-fit least-squares lines are drawn for each of the individual processes. They show that the unit event dimensions or the height to length ratio varies between processes. This can be represented by the equation:

$$Rz = A.(RSm)^{B}$$

where A and B are constants. As a first order approximation, one can say from the figure that the 'B' values are fairly constant whereas the 'A' value varies for each process. The largest Rz/RSm ratios correspond to the abrasive unit event processes like grinding and lapping and the smallest ratios correspond to cutting processes like turning and milling. Furthermore, the former processes tend to produce lower surface roughnesses than the latter.

References and further reading

Dagnall H, *Exploring Surface Texture*, Taylor Hobson Ltd, 1998.

Griffiths B J, *Manufacturing Surface Technology*, Penton Press, 2001.

ISO 1302:2001, *Indication of Surface Texture in Technical Product Documentation*, 2001.

ISO 3274:1996, *Surface Texture: Profile Method – Nominal Characteristics of Contact (Stylus) Instruments*, 1996.

ISO 4287:1997, *Surface Texture: Profile Method – Terms, Definitions and Surface Texture Parameters*, 1997.

ISO 4287:2000, *Geometric Product Specification (GPS) Surface Texture: Profile Method – Terms, Definitions and Surface Texture Parameters*, 2000.

ISO 4288:1996, *Surface Texture: Profile Method – Rules and Procedures for the Assessment of Surface Texture*, 1996.

ISO 11562:1996, *Surface Texture: Profile Method – Metrological Characteristics of Phase Correct Filters*, 1996.

ISO 12085:1996, *Surface Texture: Profile Method – Motif Parameters*, 1996.

ISO 13565–1:1996, *Surface Texture: Profile Method – Surfaces having Stratified Functional Properties, Part 1, Filtering and General Measurement Conditions*, 1996.

ISO 13565–2:1996, *Surface Texture: Profile Method – Surfaces having Stratified Functional Properties, Part 2, Height Characterisation using the Linear Material Ratio Curve*, 1996. (ISO 16610.)

Mummery L, *Surface Texture Analysis – The Handbook*, Hommelwerke Ltd, 1990.

Stout K J, Matthia T, Sullivan P J, Dong W P, Mainsah E, Luo N and Zahouani H, *The Development of Methods for the Characterisation of Roughness in Three Dimensions*, Report EUR 15178 EN, EC Brussels, ISBN 0704413132, 1993.

Stout K J, Matthia T, Sullivan P J, Dong W P, Mainsah E, Luo N and Zahouani H, *Development of Methods for the Characterisation of Roughness in Three Dimensions*, Report EUR 15178 EN, EC Brussels, ISBN 0704413132, revised edition published by Penton Press, London, 2000.

Whitehouse D J, 'The Parameter Rash – Is There a Cure?', *WEAR*, volume 83, pp 75–78, 1982.

Appendix: Typical Examination Questions

Chapter 1

1. True or false? Answers can be found in the text or in the figures in Chapter 1.
   - The correct ISO term for engineering drawing is '*Technical Product Documentation*'.
   - Engineering drawing depends upon the English language.
   - Visualization is all-important in engineering drawing.
   - The 'highest' standards are the ISO standards.
   - A grid reference system should be included on all engineering drawings.
   - There should be a 15mm border around all drawings.
   - Engineering drawings produced on a CAD system are more valid than manual (hand-drawn) drawings.
   - Noise can never enter the design process.
   - Specification can be achieved in 3D engineering drawings.
   - The preferred engineering drawing paper sizes are the 'A' series.
2. Explain why engineering drawing can be described as a language. Use any engineering drawing of your choice to illustrate your points (Sections 1.4 to 1.6).
3. Compare and contrast the following terms: '*Representation*', '*visualization*' and '*specification*' (Section 1.5).
4. Design your own engineering drawing template that you can use at any time in the future. It should include border, title

block, centring marks and whatever else you want to include from Section 1.6.

5. Explain the difference between '*computer aided draughting*' and '*computer aided design*' (Section 1.7).
6. Explain why the ISO recommend the term '*technical product documentation*' rather than '*engineering drawing*'.
7. You are the designer of the hand vice shown in Figure 1.11. You want it made and have decided to subcontract it. What types of drawings do you think you would produce to be sent to the subcontractor (Section 1.6.2)? How many of each type would you need to send to the subcontractor to SPECIFY the vice design?
8. Explain how engineering drawing prevents optical illusions (Section 1.4).
9. A subcontractor receives a set of engineering drawings from a contractor, which give details of a complicated assembly and its various parts. They are asked to manufacture all the parts and assemble the artefact. What size do you think the drawings would be and what things would be printed as standard on each? Would any 'standard' thing be on one drawing and not on another? (Section 1.6.1.)
10. Explain why it is advantageous for an engineering design company to conform to ISO standards rather than any particular national standard.

Chapter 2

11. True or false? Answers will be found in the text or in the figures in Chapter 2.

    - 3D engineering drawings should always be completed in perspective projection.
    - Axonometric projection is a particular type of isometric projection.
    - The best pictorial projection is isometric projection.
    - Cavalier projection is to be preferred to Cabinet projection.
    - Third angle projection is to be preferred to first angle.
    - Projection lines need to be included on engineering drawings.

- A sectional view of a part should always be used when there are internal details.
- There should always be at least three views of a part.
- The letters 'RSV' refer to '*reverse standard view*'.
- Second angle projection is valid under some circumstances.

12. Draw a 3D pictorial drawing of a 'block' house of your choice. For example, the roof can be a triangular block, the walls and doors can be rectangles and the windows and chimney can be squares. Avoid the use of curves.
Draw a third angle projection of your house. From this, draw a perspective projection of the house using two vanishing points.
13. Reproduce Figure 2.6 (a cube with circles on each face) in isometric projection as shown, then draw it in oblique projection.
14. Figure Q14 shows the bearing block in Figures 2.5 and 2.8. It is drawn in oblique projection and a scaled grid is included for dimensional guidance. Draw isometric as well as oblique views of the bearing block but use different viewing directions (your choice) from those in Figures 2.5 or 2.8.

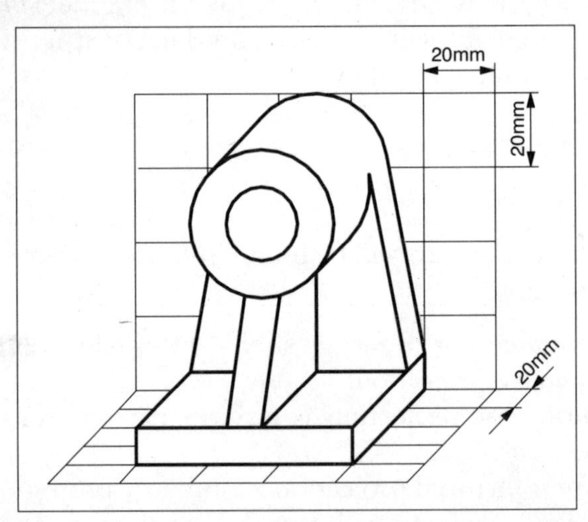

Figure Q14

15. Figure Q14 shows the bearing block in Figure 2.8 with a scaled grid for dimensional guidance. From this, draw the following views in third angle projection:

- a front view;
- a left-side view;
- a right-hand side sectional view;
- a rear view;
- a plan view;
- an inverted plan view.

Include hidden details. Do not dimension. Label the views.

16. In the sketches in Figure Q16, two drawings of various rectangular blocks are given in third angle projection. They are in the ratio one unit high and two units long. Complete the third view and then draw each in isometric as well as oblique projection. Use any convenient scale of your choice.

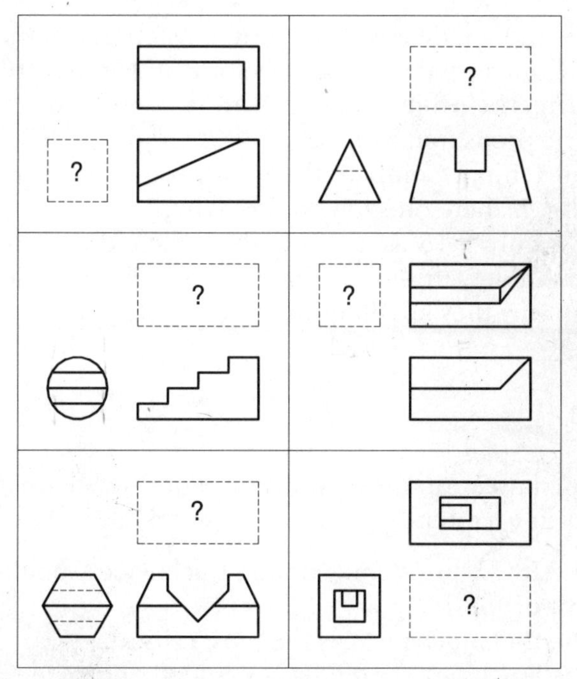

Figure Q16

17. Figure 1.12 is the drawing of the movable jaw. Redraw this in third angle projection using four views as follows:

- the front view (as shown);
- the left-hand side view section through the centre (as shown);

- a plan view;
- an inverted plan view.

Include all hidden details so that you overcome the need to have the stepped section. Do not dimension. Label the views.

18. Figure 2.16 shows the drawing of a flange. If the outside diameter is 150mm, then, using scaled measurements, draw the following views in third angle projection:

- a front view (as shown but unsectioned);
- a full plan view rather than the half plan shown;
- an inverted plan view;
- a right-hand side sectional view projected from the front view.

Include hidden details. Do not dimension. Label the views.

19. Choose one of the rectangular blocks in Figure Q16 and draw it in trimetric projection with, say, $\alpha = 40°$ and $\beta = 10°$, and dimetric projection with, say, $\alpha = 20°$ and $\beta = 20°$. Ignore any foreshortening. Compare these with your isometric projection drawing. Is there one you prefer? Why?
20. Using Figure 2.15 as a guide, draw second and fourth angle projection drawings of the block shown. From these drawings, explain why they are illogical projections.

Chapter 3

21. True or false? All answers will be found in the text or in the figures in Chapter 3.

- The ISO type 'A' and 'B' line thicknesses should be in the proportion 1:2.
- The ISO line type 'A' is the most critical.
- The line types 'C' and 'D' are interchangeable.
- Cross hatch lines are at 45° wherever possible.
- Sections are always cross hatched, irrespective of the size or length of the section.
- It is not necessary to have a terminator at the end of a leader line.
- Dimension projection lines do not always have to be type 'B' lines.

- The ISO recommended decimal marker is a comma.
- The Greek letter 'ϕ' must always be used to indicate diameter.
- Flat surfaces such as squares, tapered squares can be represented in their side view by a '+' sign.
- When drawing splines or gears, each and every tooth needs to be included in the drawing.
- Colour is not recommended in engineering drawings.

22. Using your intuition, guesstimate the ranking of the 10 ISO line types in Figure 3.4 according to the frequency of their use in engineering drawings in general. To help you, I think type 'A' is used the most because it is the principal line for part outlines and shapes. I think type 'B' is a very close second because it is used for cross-hatching and dimensions. What do you think about my thoughts and about the other line types? Would you expect the ranking to be different for detailed drawings as opposed to assembly drawings? (Section 3.2.)
23. With respect to the movable jaw drawing in Figure 3.2, count the number of lines in each of the 10 ISO line type classes. Work out the percentages of each and from this, rank the 10 according to their frequency of use. Compare your answer with your guesstimate. (Section 3.2.)
24. With respect to the assembly drawing in Figure 3.1, count the number of lines on the drawing in each of the 10 ISO line type classes. Work out the percentages of each and hence determine the ranking of the frequency of use. Compare this answer with your guesstimate. (Section 3.2.)
25. Draw a section through a threaded bolt located in a threaded hole. The male threaded bolt should not be sectioned but the hole should be. The reason for this question is to ensure you understand the use of the line types A and B for male and female thread forms. (Section 3.8.3 and Figures 3.5 and 3.6.)
26. Using your template from Question 4, redraw the vice assembly drawing in Figure 3.1 in third angle projection but include the following views:

    - a full sectional front view (rather than the partial front view shown);
    - a plan view;
    - a left-side view (as shown);
    - a right-side view.

Include hidden detail as appropriate. Add a balloon reference system. Add an item list.

27. Using your template from Question 4, redraw Figure 3.3 of the hardened insert in third angle projection but without the dependency on the symmetry as given by the 'equals' sign. Add dimensions sufficient for it to be made. This could involve adding dimensions from the sides. Beware of the error of repeating dimensions. Include at least one auxiliary dimension.
28. Figure 2.16 shows the drawing of a flange. Assume the outside diameter is 150mm. Reproduce the two drawings as shown in third angle projection and, using scaling measurements, add dimensions sufficient for the flange to be manufactured. Use your template from Question 4.
29. Using your drawing template from Question 4, redraw in third angle projection the movable jaw in Figure 3.2 but draw the following three views:

    - the front view (as shown);
    - the left-hand side view section through the centre;
    - a right-hand side section through one of the ϕ5/ϕ8 counter-bored holes.

    Fully dimension the jaw sufficient for it to be manufactured and include hidden detail if you think it helps understanding. Add the title information.
30. With reference to Figure Q30, draw the nut, bolt and washer assembly full size for M20 as well as separate detail drawings of the nut, bolt and washer. Use third angle projection. The bolt length should be 60mm and the thread length 40mm. Dimension the detail drawings. Use your drawing template from Question 4.

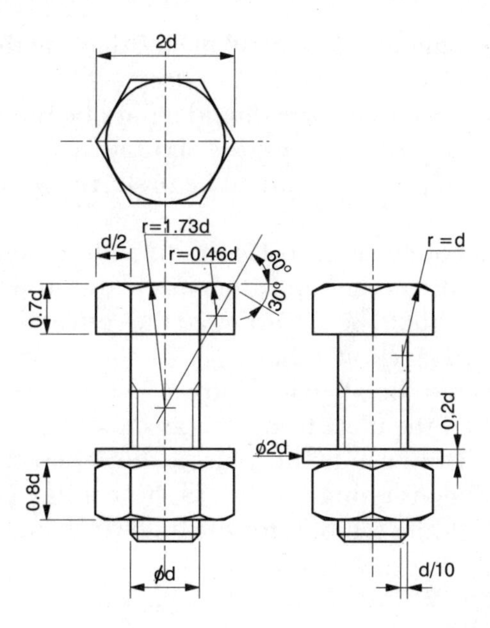

Figure Q30

Chapter 4

31. True or false? All answers can be found in the text or in the figures in Chapter 4.

- There is no such thing as a 'non-functional dimension' since all dimensions are functional.
- It doesn't matter if dimensions are given twice on a drawing.
- There are six elements to any dimension.
- It doesn't matter if dimension lines are crossed or separated by other lines.
- With respect to dimensioning angles, it is common to have only one terminator.
- It doesn't matter if the units used for the dimension value are different to the units used for the associated tolerance.
- All dimension values, graphical symbols and annotations should be added to a drawing such that it can always be read from the bottom and the right-hand side.
- Chain dimensioning should always be used wherever possible.
- Projection lines should always touch the outside outline of a part.

- The designation 'M8 8/10' means that the threaded 8mm diameter hole is to be 0.8mm deep.
- The smallest tolerance should always be used because this means the part is always very accurate.
- Unilateral tolerances are to be preferred to bilateral tolerances.

32. Draw the stepped shaft in Figure 4.3, add an end view using third angle projection and dimension the drawing correctly. Dimensions should be according to the convention in the latest ISO standard (Section 4.2).
33. With respect to the plate in Figure 4.7, if the maximum height is 10cm, draw the plate front view as shown using scaled measurements. Dimension the plate using correct rather than incorrect dimensioning practice (Section 4.2).
34. Draw the welding symbols for the following situations (Section 4.3):
    a A single V butt weld with a backing run produced by manual metal-arc welding, reference number 111 in ISO 4063. The backing weld is to be ground flat.
    b A spot seam weld, 5mm wide, consisting of 3 runs of 10mm length with 10mm between them.
    c A 'T' shaped fillet weld with fillet welds on with side.
35. Using the data in Figure 4.13, draw a graph of accuracy against cost. What do you conclude from this? Is the answer clearer if a log-log graph is used?
36. Identify the datum features, the functional and the non-functional dimensions of the movable jaw shown in Figure 3.2 and the hardened insert in Figure 3.3.
37. With reference to the table in Figure 4.13, the worst case is drilling which, if one simplistically adds the 36um and 14um errors together, makes almost 50um. Bearing in mind this applies to a 25mm diameter hole, it is equivalent to 0,2%. As a first order approach, one could use this proportion to define tolerance values. This 0,2% is the error one can expect from a machining process and we should add a 'factor of safety' if we are to translate it to a tolerance. Let's take a factor of, say, 5, which makes 1,0%. Thus, if we assumed dimensions and tolerances are linearly related, we could use the equation to calculate bilateral tolerance values:

'BILATERAL TOLERANCE = ±(SIZE 0,5%)'

Using this equation, add tolerances to your drawings from Questions 32 and 33. To help you, a 30mm diameter would be: 'ϕ30 ± 0,15'.

38. Using the hole symbology of Section 4.3, draw the following holes (like the drawings in Figure 4.8) and add dimensions:

For 'thick' sections:	ϕ11 × 10	ϕ22 × 3U
	M10 × 35/42	ϕ11 × 15V
For a 30mm thick plate:	ϕ20 × 13U	□20
	ϕ13,5	

In each case draw a section through the hole as well as a plan view (Section 4.3 and Figure 4.8).

39. Add the general tolerance ranges given in Section 4.5 (ie XX and XX,X and XX,XXX) to your template from Question 4 so that you can always refer to general tolerance values when you draw in future. You don't have to use the values I suggest. Why not explore what is said in other books and publications?
40. In the Figure 4.16 example, there are no out-of-roundness errors. Hence, the maximum and minimum hole centre spacings are 22,00 and 23,00 (as shown). In reality there must be some out-of-roundness, as shown in Figure 4.17. If the bolts and holes have out-of-roundness' of ±0.5% (Question 37 above), calculate the new maximum and minimum hole centre spacings which still permit assembly. There will be two sets of values corresponding to the ϕ5 bolt shank in its ϕ5,5 hole and the ϕ8 bolt head in its ϕ8,5 hole. Which one is the governing case?

Chapter 5

41. True or false? All answers can be found in the text or in the figures in Chapter 5.

    - The clearance in a '*close-running fit*' is smaller than that in a '*sliding fit*'.
    - There is no out-of-roundness in a hole drilled by a new sharp drill.
    - The IT5 tolerance range is larger than the IT4 tolerance range.

- One of the values of the 'H' or 'h' tolerance classes is always zero.
- The tolerance class c11 is the negative of class C11.
- The GT symbol for symmetry is an 'equals' sign.
- A datum must always be given in a GT box.
- Reaming produces a hole with more variability than honing.
- The relationship between the IT tolerance range and nominal size is linear.
- The fit classes A to Z are a linear series.
- Clearance fits are always associated with classes A to H (or a to h).
- The shaft basis system of fits is superior to the hole basis system.

42. Explain the meaning of the following terms: limits, fits, universal tolerance, specific tolerance, IT tolerance grades, geometrical tolerance (Sections 4.5, 5.1, 5.3, 5.4 and 5.6).
43. The 9,1um out-of-roundness produced by the worn ϕ10mm drill in Figure 5.2 corresponds to tolerance range IT7 (Section 5.3). If the diameter was different, the 9,1um error would correspond to a different class. Determine the IT class the 9,1um corresponds to for ϕ5, ϕ10, ϕ15, ϕ20, ϕ25, ϕ30, ϕ35, ϕ40, ϕ45 and ϕ50mm holes. Plot the results in graphical form.
44. The table in Figure 4.13 gives errors in microns for ϕ25mm holes produced by various manufacturing processes. Translate these into IT equivalents. (Note that the average roughness Ra cannot translate into an IT equivalent.) The taper, ovality and roundness are individual errors that contribute to an overall error. How do the Figure 4.13 individual errors sum to an overall error value which can be translated into an IT range? What is the overall IT value for each process?
45. Give the shaft and hole dimensions for the following tolerance cases. In each case, state the maximum, minimum and average clearance/interference values (Figures 5.11 and 5.12):

    ϕ15 G7/h6; ϕ100,00 H7/n6; 37,5 h6/S7
    □22,1 F8/h7; 10,00 G7/h8; ϕ25,4 H6/g6

46. Translate Figure 5.4 into a graph, i.e. plot a family of curves of IT1, IT2, IT3 ... IT11 against nominal size. Use the average of the nominal size range. This means that for the nominal size class of 120 to 180, you should use the average of 150 to plot

the graphical values. Are the relationships linear, parabolic or what? Add to your graph the linear relationship ±(SIZE 0,5%) from Question 37. Comment on the various curves and their relationships. Is the Question 37 equation valid?

47. Add another column to the table in Figure 5.6 that states whether the process produces flat surfaces or cylindrical surfaces. Note that some processes can be used for either (e.g. lapping). From this, select two processes that can be used to machine the surfaces appropriate to each of the following fit classes. In each case select the process which gives minimum cost. Assume that the processing cost increases as the accuracy increases.

    - Flat: Loose-running fit – H11/c11 (Figure 5.11).
    - Cylindrical: Sliding fit – H7/g6 (Figure 5.11).

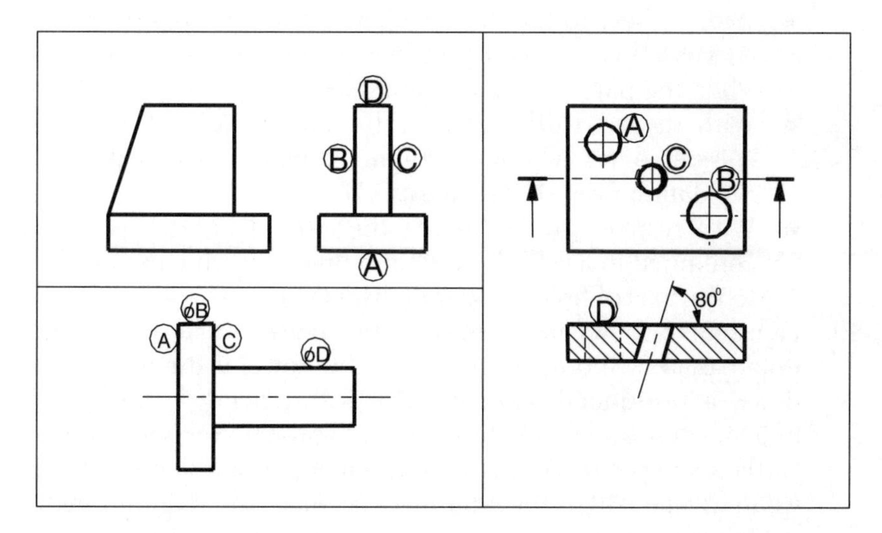

Figure Q48

48. With reference to the third angle projection drawings in Figure Q48, reproduce each figure but include a geometric tolerance box for the following cases (Sections 5.5 and 5.6):

    - With respect to the 'tee' piece, face 'B' is to lie between two parallel planes 0,15mm apart that are perpendicular to datum face 'A' (perpendicularly GT).

- With respect to the 'tee' piece, face 'C' is to lie between two planes 0,10mm apart that are parallel to datum face 'B' (parallelism GT).
- With respect to the 'tee' piece, the top face 'D' is to lie between two parallel planes 0,02mm apart (flatness GT).
- With respect to the 'turned' part, the axis of diameter 'D' is to lie in a cylindrical tolerance zone 0,1mm diameter (straightness GT).
- With respect to the 'turned' part, the curved surface of diameter 'B' is to lie between two co-axial cylindrical surfaces 0,02mm radially apart (cylindricity GT).
- With respect to the 'turned' part, the axis of diameter 'D' is to be contained in a cylindrical tolerance zone 0,05mm diameter which is co-axial with the axis of diameter 'B' (concentricity GT).
- With respect to the 'turned' part, at the periphery of any cross-section, runout of surface 'B' is not to exceed 0,1mm when the part is rotated about diameter 'D' (runout GT).
- With respect to the 'plate', the axis of hole 'B' is to be contained in a cylinder 0,06mm diameter inclined at 80° to the datum face 'D' (angularity GT).
- With respect to the 'plate', the axis of hole 'A' is to be contained in a cylinder 0,03mm diameter which is parallel to the axis of hole 'C' (parallelism GT).

49. Figure 5.15 shows an assembly of two plates and a dowel. The upper plate will only assemble on the dowel if the top of the dowel is positioned correctly. This will depend on the dowel hole position and inclination (assuming the dowel is not bent!). In the example in Figure 5.15, the inclination tolerance zone value was ϕ0,030 as an example. Calculate the required inclination tolerance zone value for the worst case situation, i.e., the upper plate hole and dowel diameters are at the worst extremes of their H11 and C11 ranges. Assume the upper plate hole is exactly on centre.
50. With reference to the movable jaw drawing in Figure 3.2, there are two functional 'bearing' surfaces. These are the 16mm wide tongue which locates in the body and the $\phi 15 \times 7{,}5$ central hole in which the bush (part 4) rotates. Using the fits tables in Figures 5.11 and/or 5.12, select fits appropriate to these two situations. From this, add these tolerances as well as general

tolerances to your movable jaw detail drawing from Question 29.

Chapter 6

51. True or false? All answers can be found in the text or in the figures in Chapter 6.

    - All surfaces contain short and long wavelength components.
    - The evaluation length is five times the sample length.
    - The parameter 'Rq3' is the RMS value of the third sample length.
    - The parameter 'Ra' is a spacing parameter.
    - The parameter 'Ra' is the most important parameter.
    - The 16% rule says the surface is considered acceptable if more than 16% of the measured values are less than the value specified.
    - Only the 'Ra' value can be shown at position 'X' on the tick symbol.
    - The 'point' of the tick symbol should be placed on the surface outline or an extension to it.
    - It is important to specify the surface finish of each and every surface of a part.
    - The surface lay symbol 'C' means the surface must be cut.
52. Using a surface finish parameter of your choice, explain the formula '*TnN*' (Section 6.3).
53. Obtain a set of random numbers. Assume them to be in microns. Assume that they are the sample length data points. Using this set:

    - calculate the equivalents of Rz, Ra, Rp and Rv values;
    - check that the skew parameter (Rsk) value is zero for the random set.
54. Explain the meaning of the following terms: '*sample length*' and '*evaluation length*' (Section 6.2).
55. If the evaluation length peak to valley height of the schematic profile in Figure 6.9 is 10 microns, use scaled measurements to calculate:

- the five sample length Rz values;
- the five sample length Rv values;
- the five sample length Rp values.

56. Explain why it is necessary to use a filter on a set of raw surface finish data before calculating any of the roughness parameters (Section 6.2).
57. Explain the meaning of the following terms: BAC, ADF (or HDF), PnN, m_3, PC filter, Rmr(c).
58. Draw the ISO 1302 (or BS 308) 'tick' symbol for the following conditions (Section 6.5):

    - The specified surface is to be polished such that, when it is measured using a profilometer set for a sampling length of 0,25 mm, the Rz value must be less than 1,0μm.
    - The specified surface is to be lapped such that the surface finish is between 0,2 and 0,4um Ra when the sampling length is 0,25 mm.
    - The specified surface is to have a maximum surface finish of Rz = 3,0um using the 16% rule. The surface is to be ground such that the lay is perpendicular to the plane of the drawing. All other values are to be the default ones.
    - The specified surface is to be cast and no machining is permitted after casting. The surface finish must be no greater than Rq = 1,0um.

59. Sometimes it is beneficial to use average SF parameters like Ra in preference to extreme parameters like Rz. As a first order approximation, one can say that 9Ra = Rz. Using this value, convert the various Rz values given in Figure 6.18 to Ra values and create a new process capability table of Ra against manufacturing process similar to the one in Figure 5.6.
60. Figure 6.16 shows the lay classes according to ISO 1302:2001. Research the lay produced by manufacturing processes and add another column to the table in Figure 5.6 stating the lay class.

General

61. Figures 3.2 and 3.3 are detail drawings of the movable jaw and the hardened inserts respectively. Using the dimensional information in these figures and scaled measurements from the

assembly drawing in Figure 3.1, draw detail drawings in third angle projection of the other parts. Include dimensions and tolerances. Also include geometric tolerances and surface finish specifications where you think appropriate. These are the body (part 1), the bush (part 4), the bush screw (part 5), the jaw clamp screw (part 6), the tommy bar (part 7), the plate (part 9) and the plate screw (part 10). Provide an item list detailing all the parts but which also gives information about the insert screws (part 8).

62. Using your drawing template from Question 39, reproduce the pulley system detail and assembly drawings in Figure 4.1 in third angle projection. The bolt bearing diameter is to be 20mm. The thread is to be M15. The bearing fit is to be a free-running fit. Using scaled measurements, draw detail drawings of the shaft, pulley and hole (local section), an assembly drawing and an item list of parts in third angle projection. Your drawings should include dimensions and tolerances (universal or specific as appropriate) sufficient for the system to be made by a subcontractor in another country. Using the information in Figure 5.6, select manufacturing processes for the bolt bearing and the pulley hole. The shaft and bolt material is mild steel. Add a GT for the two end faces of the pulley.
63. The photographs in Figure Q63 show an engineer's clamp that has been made in imperial units. The jaws are 9/16 inches square and the paper used for the background is 1cm graph paper. Using scaled measurements, convert the dimensions to the nearest logical metric units and draw an assembly drawing, detail parts drawings and an item list in third angle projection. Add dimensions and tolerances sufficient for it to be made. The materials of construction are steel.

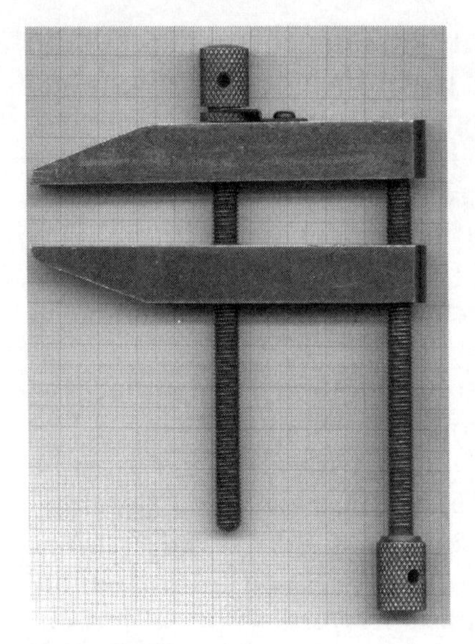

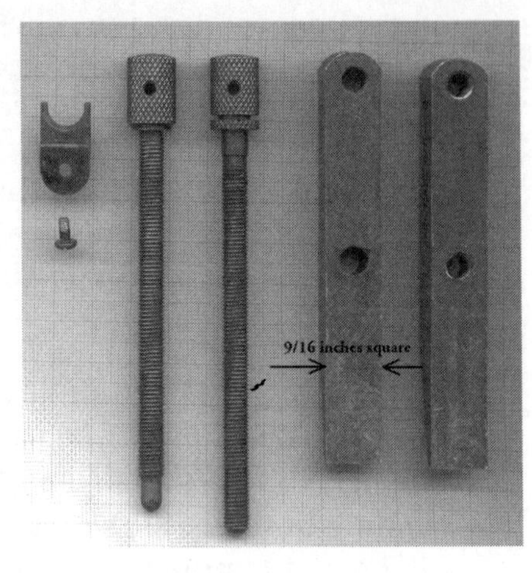

Figure Q63

64. The photographs in Figure Q64 show a woodworking adjustable bevel which was made in imperial units. The photo shows a ruler for scaling purposes. The background is 1cm graph paper. Note that the parts photo shows four parts whereas there are really nine parts, i.e. those shown plus four rivets and the spacer plate. Using scaled measurements, convert the dimensions to the nearest logical metric units and draw an assembly drawing, detail parts drawings and an item list in third angle projection. Add dimensions and tolerances sufficient for it to be made. The materials of construction are steel. The blade is 0,067 inches thick gauge plate steel. The two sides are riveted together using four double-sided rivets. I suggest you use ϕ3mm countersunk rivets with head angles of 90°. Standard rivets have maximum head diameters after forming of 1,85 shank diameter (i.e. 5,55mm). This means that, using the symbology in Section 4.3 and Figure 4.8, each side of the rivet holes is given by:

$\phi 5{,}55 \times 90°$
$\phi 3$

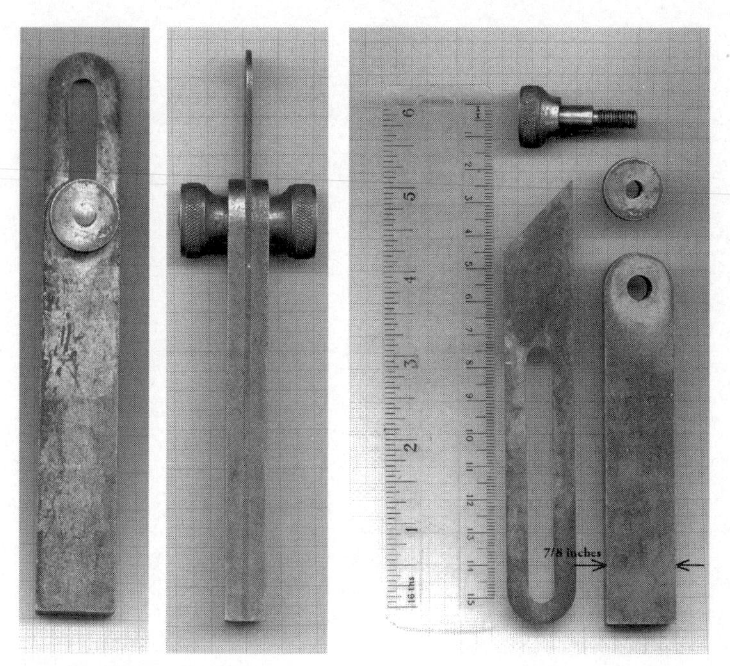

Figure Q64

65. Figure Q65 shows a 'site' sketch of a flange used for joining pipes that convey high-pressure liquid. It gives only basic information and the intention is that the designer will later produce full engineering drawings. The intention is that two pipes will be joined using two flange/pipe assemblies, a 1mm thick PTFE gasket (ID = 35mm) and the necessary bolts, washers and nuts. Draw a full assembly drawing, an item list, a detail drawing of the gasket, a detail drawing of the flange (prior to welding), a detailed drawing of a pipe end prior to welding and a flange + pipe welded assembly drawing in third angle projection. Dimension and tolerance the drawings sufficiently for the parts to be made.

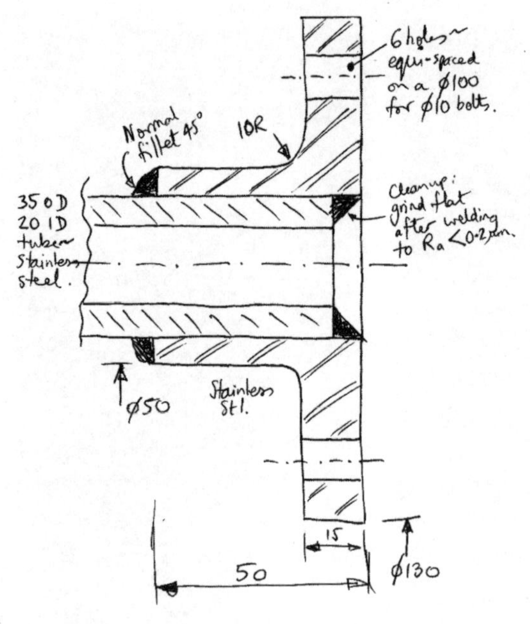

Figure Q65

66. Obtain a component that is simple and commonly available and produce a detail drawing of it sufficient for it to be manufactured. Such a component could be: a paperclip, key, drawing pin, ruler, centre punch, coat hook, glass jar, special nut or washer or bolt (e.g. casellated, lock), spanner, nail, paper cup, plastic cup, CD, needle, cotton reel, cable tie, house brick, cardboard cereal box.
67. Beg, borrow or buy an artefact that consists of an assembly of parts and perform a reverse engineering exercise on it by analysing the constituent parts, making measurements, working out how the parts were made, the materials, the processes and how they are assembled. From this analysis, draw an assembly drawing of the artefact, detail drawings of the parts and an item list in third angle projection. Include dimensions and tolerances sufficient for it to be made.

    - Typical candidate small-scale assemblies are: a bicycle pump, torch, pencil sharpener, floppy disk, audio tape, audio tape box, CD box, fizzy drink can, craft knife, electrical plug or socket, door lock, stapler, paint brush, biro, pencil, pipe clip, hole centre punch, highlighter, door handle, 'Sellotape' dispenser.

- Typical candidate large-scale assemblies are: a desk, chair, stool, bench, bookcase, door, window, shelving system, ducting, rainwater piping, cardboard box and its packaging, road sign, manhole cover, picture frame, filing tray set, lamp, car jack, TV aerial, stepladder, flat-pack furniture.

68. The drawing in Figure Q68 shows a section through a relief valve assembly. The spring length within the assembly is 52,00mm and it has a rate of 1.05kg/mm. The critical dimensions are as follows: a = 58,00 ± 0,1; b = 3,00 ± 0,1; c = 18,00 ± 0,1 and d = 2,00 ± 0,05. The body, cap and piston are made of steel and the washer of aluminium alloy. Using scaled measurements, draw detail drawings, an assembly drawing and an item list in third angle projection. Add dimensions and tolerances sufficient for it to be made.
69. A company makes a hinge for wooden doors, see Figure Q69(a). The hinge consists of three parts: two identical drilled and folded strips and a pin. The flat part of the strip is 75mm wide,

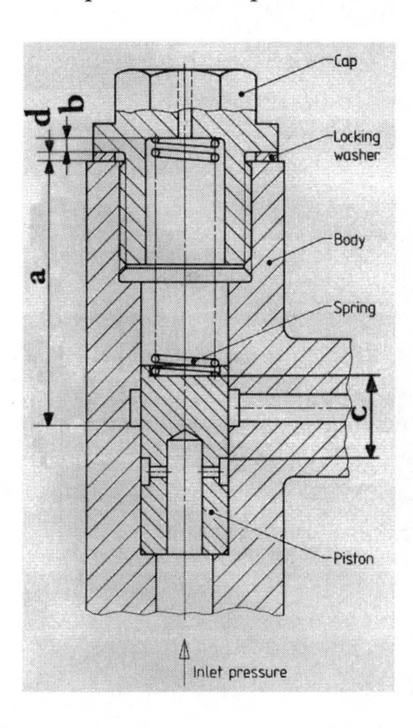

Figure Q68

20mm wide and 1,5mm thick mild steel. The holes for the screws are 5,5mm diameter. The hole centre distances are as shown in the figure.

The sketch in Figure Q69(b) is a designers 'ideas' sketch for the small hand-held jig to be used for drilling the wood-screw holes in the hinge strips. The jig is to consist of a base body and a top plate. The top plate needs to be screwed to the base body so that it can be removed and replaced by another plate having a different arrangement of holes. The top plate needs to have three holes in it for drill bushes, four holes for clamping it to the base and hole/s for some form of clamping bolt/s to hold the hinge during drilling. The jig is to be made from medium carbon steel. The clamping bolt/s need only be simple hexagonal headed ones. A drawing of the company's standard 5,5mm-drill bush is shown in Figure Q69(c).

Design the jig (i.e. you decide the bolt sizes and the jig slot size to receive the strip, etc.) and hence draw an assembly drawing, detail drawings and an item list of the jig in third angle projection. Add dimensions and tolerances sufficient for it to be made.

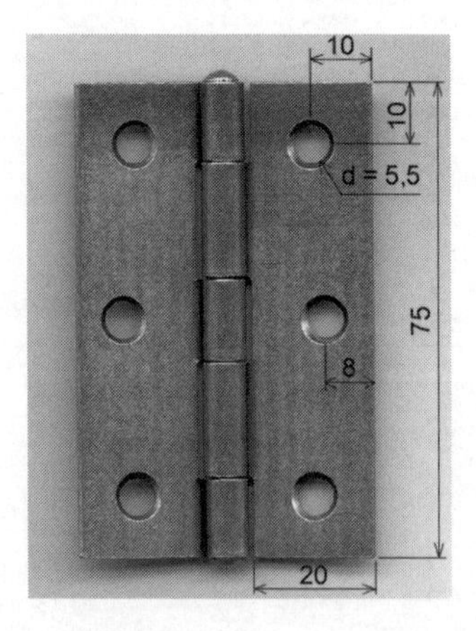

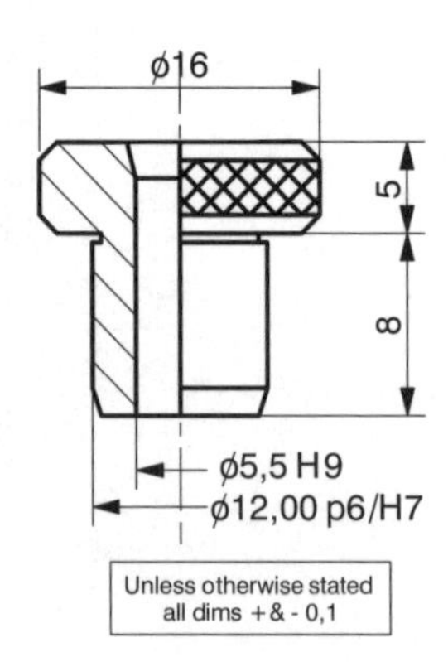

Figure Q69(a), (b)

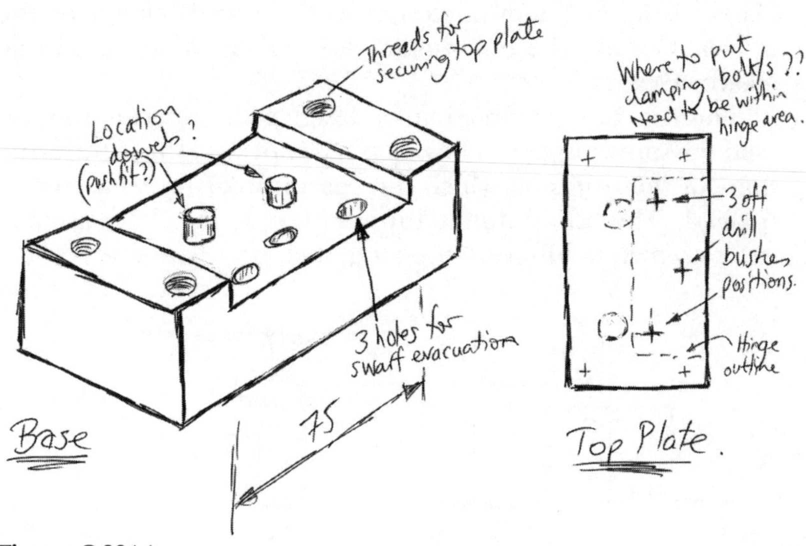

Figure Q69(c)

70. The following exercise has been used for a number of years at Brunel University as the 'design and make' project. It involves the design and manufacture of a set of weighing scales. Every student designs and makes his or her own set.

    The scales comprise of four basic sets of parts: a weighing sensor, electronic circuit, support frame and pan. The sensor assembly is shown in Figure Q70(a), the circuit assembly in Figure Q70(b) and the item list in Figure Q70(c). Note, Figure Q70(b) shows the position of the LED's on the front face of the PCB when their output values are read through the Support Plate. The two Leaf Springs (Figure Q70(a)) to which strain gauges are cemented, are made of spring steel. Each has two 5.2mm holes drilled at the ends. These holes are for the M5 bolts clamping them to the Block at one end and the M5 Stud clamping them to the Spacer at the other. Above the Spacer is the Pan Base on top of which is the Pan. M5 bolts clamp the sensor assembly to the Support Plate. The electronic circuit panel (Figure Q70(b)) is attached to the Support Plate via adhesive pads on the PCB support pillars that need to be positioned such that the two LED's can be read through a cutout. A battery is attached to the Support Plate, via an adhesive pad. Two switches (on/off and reset), the connecting wires, the strain gauges and the electronic circuit complete the construction.

There is no freedom of design with the PCB circuit or the sensor because the unit would not work if these were to be changed.

There is complete freedom of design with respect to the pan and the support plate. The pan is made of steel sheet and is the part of the scales on which the bag of sugar (or whatever) is placed. The aluminium support plate is the 'body' of the scales, which holds everything together.

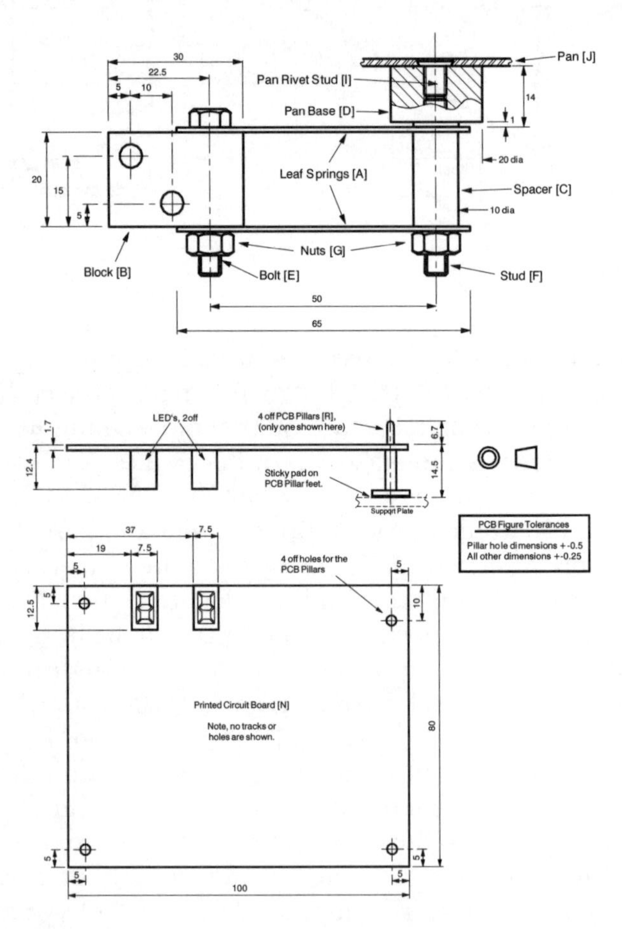

Figure Q70(a), (b)

Part	Component	No	Material	Operation	Size/Length [mm]
A	Leaf Springs	2	Steel	Supplied	65x10x0.7 with holes
B	Block	1	Alum	M/ced	Raw material is 22.3 x 12.7 plate. Student to cut off a 32mm length from the plate and mill to size. Final size to be 20 x 30 x 12.7 with N5.2 holes
C	Spacer	1	Steel	M/ced	Raw material is N11.1 steel bar. Student to cut off a 22mm length from the bar and turn to size. Final size to be N10 x 20 long (plus N5.2 central hole).
D	Pan Base	1	Brass	M/ced	Raw material is N23.7 brass. Student to cut off a 15mm length from the bar and turn to size. Final size to be N23.7 x 14 with a 1mm step and a M5 hole.
E	Bolt M5 x 30	1	Steel	Supplied	
F	Stud M5	1	Steel	Supplied	Cut to length
G	Nuts M5	2	Steel	Supplied	
H	Strain Gauges	2		Supplied	Cemented to Leaf Springs
I	Pan Stud Rivet	1	Steel	Supplied	M5x8. Punched into Pan through the N5.1 hole
J	Pan	1	Steel	Cut/form	Raw material is 0.91mm (20 gauge) x 130 x 130 steel sheet. Student to decide final size and proportions but no less than 960x60. Rivet hole drill 5.1
K	Support Plate	1	Alum	Cut/form	Raw material is 1.2mm (18 gauge) x 150 x 250 aluminium sheet. Student to decide final size.
L	Bolts M5 x 25	2	Steel	Supplied	(For bolting Block to Support Plate)
M	Nuts M5	2	Steel	Supplied	(For bolting Block to Support Plate)
N	Electronic Circuit	1	PCB & compts	Assemble & solder	Components soldered to PCB. Final PCB size is 1.7 x 100 x 80.
O	Wire to Gauges	2	Wire	Cut	Cut to length
P	Switch	1		Supplied	2 off N6.5 holes needed in the Support Plate for mounting holes, 15mm clearance needed behind and 17 in front
Q	Reset Button	1		Supplied	
R	PCB Pillars	4	Plastic	Supplied	Insert into PCB & stick onto Support Plate

Figure Q70(c)

Background and Rationale of the Series

This new series has been produced to meet the new and changing needs of students and staff in the Higher Education sector caused by firstly, the introduction of 15 week semester modules and, secondly, the need for students to pay fees.

With the introduction of semesters, the 'focus' has shifted to module examinations rather than end of year examinations. Typically, within each semester a student takes six modules. Each module is self-contained and is examined/assessed such that on completion a student is awarded 10 credits. This results in 60 credits per semester, 120 credits per year (or level to use the new parlance) and 360 credits per honours degree. Each module is timetabled for three hours per week. Each semester module consists of 12 teaching weeks, one revision week and two examination weeks. Thus, students concentrate on the 12 weeks and adopt a compartmentalized approach to studying.

Students are now registered on modules and have to pay for their degree per module. Most now work to make ends meet and many end up with a degree and debts. They are 'poor' and unwilling to pay £50 for a module textbook when only a third or half of it is relevant.

These two things mean that the average student is no longer willing or able to buy traditional academic text books which are often written more for the ego of the writer than the needs of students. This series of books addresses these issues. Each book in the series is short, affordable and directly related to a 12 week teaching module. Each will only be six chapters long, giving one chapter per two teaching weeks. Thus, module material will be presented in an accessible and relevant manner. Typical examination questions will also be included which will assist staff and students.

However, there is another objective to this book series. Because the material presented in each book represents the state-of-the-art practice, it will also be of interest to industrialists and specialist practitioners. Thus, the books can be used by industrialists as a first source reference that can lead onto more detailed publications.

Therefore, each book is not only the equivalent of a set of lecture notes but is also a resource that can sit on a shelf to be referred to in the distant future.

Index